Mathematical Models for Teaching

Mathematical Models for Teaching

Reasoning without Memorization

Ann Kajander and Tom Boland

Canadian Scholars' Press
Toronto

Mathematical Models for Teaching: Reasoning without Memorization
Ann Kajander and Tom Boland

First published in 2014 by
Canadian Scholars' Press Inc.
425 Adelaide Street West, Suite 200
Toronto, Ontario
M5V 3C1

www.cspi.org

Copyright © 2014 Ann Kajander, Tom Boland, and Canadian Scholars' Press Inc. All rights reserved. No part of this publication may be photocopied, reproduced, stored in a retrieval system, or transmitted, in any form or by any means, electronic, mechanical, or otherwise, without the written permission of Canadian Scholars' Press Inc., except for brief passages quoted for review purposes. In the case of photocopying, a licence may be obtained from Access Copyright: One Yonge Street, Suite 1900, Toronto, Ontario, M5E 1E5, (416) 868-1620, fax (416) 868-1621, toll-free 1-800-893-5777, www.accesscopyright.ca.

Every reasonable effort has been made to identify copyright holders. CSPI would be pleased to have any errors or omissions brought to its attention.

Canadian Scholars' Press Inc. gratefully acknowledges financial support for our publishing activities from the Government of Canada through the Canada Book Fund (CBF).

Library and Archives Canada Cataloguing in Publication
Kajander, Elizabeth Ann, 1960-, author   Mathematical models for teaching : reasoning without memorization / Ann Kajander and Tom Boland.

Includes bibliographical references and index. Issued in print and electronic formats.
ISBN 978-1-55130-556-1 (pbk.).–ISBN 978-1-55130-557-8 (pdf).–ISBN 978-1-55130-558-5 (epub)

1. Mathematics – Study and teaching (Elementary). I. Boland, Tom, 1959-, author II. Title.

QA135.6.K34 2014   510   C2013-907175-X   C2013-907176-8

Text design by Integra
Cover design by Em Dash Design

Printed and bound in Canada

Dedication

Ann: I would like to send this book out in spirit to Jari, a model of a dedicated teacher, a kind human being, and a good Finn. As well, I would like to dedicate the book to Bonnie, Gracie, Misty, and Taffy, and rescue dogs everywhere, particularly those who suffer terribly as breeder dogs in puppy mills, spending their lives in tiny cages producing puppies for the mass market.

Tom: I dedicate this book to all teachers who have the wisdom and the courage to reflect on their practice, to commit to lifelong learning, and to sometimes step out of their comfort zones in the interest of helping their students to succeed.

Contents

Preface		xi
Acknowledgements		xiii
Chapter 1:	Introduction to Mathematics for Teaching	1
Chapter 2:	Introduction to Mathematical Reasoning	8
	2.1 Mathematical Models	8
	2.2 Mathematical Communication	10
	2.3 From Examples to Generalizations	10
	2.4 Argument, Proof, and Rigor	11
Chapter 3:	Introduction to Numbers	14
	3.1 From Quantity to Number Symbols	14
	3.2 Quantity, Digits, and Place Value	14
	3.3 Exploring Models and Manipulatives	15
	3.4 Zero	17
	3.5 Between Zero and One (or between One and Two, or between Two and Three, and So On)	17
	3.6 Infinity	18
Chapter 4:	Whole Number Addition and Subtraction	20
	4.1 Addition: Concept and Contexts	20
	4.2 Addition Models, Strategies, and Procedures	22
	4.3 Subtraction: Concept and Contexts	24
	4.4 Subtraction Models, Strategies, and Procedures	27
	4.5 The Connection between Addition and Subtraction	29
	4.6 Connecting to Multiplication Concepts and Contexts	30
Chapter 5:	Whole Number Multiplication and Division	32
	5.1 The Connection between Addition and Multiplication	32
	5.2 Multiplication Models, Strategies, and Procedures	32
	5.3 Division: Concept and Contexts	39
	5.4 The Connection between Multiplication, Division, and Ratio	40
	5.5 Division Models, Strategies, and Procedures	43
	5.6 Factors, Multiples, Divisibility, and Prime Numbers	47
	5.7 Rethinking BEDMAS	52
	5.8 Fact Families	53
Chapter 6:	Fraction Representations and Additive Operations	57
	6.1 Fractions: Concept and Contexts	57
	6.2 Exploring Representations, Models, and Manipulatives	59
	6.3 Comparing Fractions	62
	6.4 Concrete Equivalent-Fraction Models	64
	6.5 Concrete Models and Strategies for Adding and Subtracting Fractions	65
	6.6 Construction of the Common Denominator	68
	6.7 Developing Procedures for Adding and Subtracting Fractions	71

Chapter 7: Multiplicative Fraction Representations and Operations 77
- 7.1 Fraction Multiplication: Concept and Contexts 77
- 7.2 Concrete Models and Strategies for Multiplying Fractions 80
- 7.3 Developing Procedures for Multiplying Fractions 84
- 7.4 Fraction Division: Concept and Contexts 88
- 7.5 Concrete Models and Strategies for Dividing Fractions 89
- 7.6 Models for Division by a Unit Fraction 93
- 7.7 Developing Procedures for Dividing Fractions 95
- 7.8 Fraction Conversions (Lowest Terms, Improper Fractions, and Mixed Numbers) 101

Chapter 8: Decimal and Percent Representations and Operations................. 108
- 8.1 Decimal and Percent: Concept and Contexts 108
- 8.2 Constructing Representations 109
- 8.3 Decimal and Percent Addition and Subtraction Concepts and Contexts 113
- 8.4 Concrete Models and Strategies for Adding and Subtracting Decimals 113
- 8.5 Developing Procedures for Adding and Subtracting Decimals 115
- 8.6 Multiplication and Division Concepts and Contexts, and the Important Role of Estimating 115
- 8.7 Developing Procedures for Multiplying and Dividing Decimals 117

Chapter 9: Integer Representations and Operations 122
- 9.1 Integers: Concept and Contexts 122
- 9.2 Extending the Number Line 123
- 9.3 Exploring Integer Representations, Models, and Manipulatives 124
- 9.4 Using Zero Pairs to Represent Integers 124
- 9.5 Integer Addition and Subtraction: Concept and Contexts ... 126
- 9.6 Concrete Models and Strategies for Adding and Subtracting Integers 128
- 9.7 Practicing Mentally Adding and Subtracting Integers 133
- 9.8 Integer Multiplication: Concept and Contexts 133
- 9.9 Concrete Models and Strategies for Multiplying Integers 134
- 9.10 Developing Procedures for Multiplying Integers 139
- 9.11 Developing Procedures for Dividing Integers 140

Chapter 10: Beyond Integers.. 147
- 10.1 Numbers on the Number Line... 147
- 10.2 Density: The Concept of "Between" 148
- 10.3 Formal Descriptions... 148
- 10.4 Irrational Numbers 150

Chapter 11: From Patterns to Algebra 153
    11.1 Designs and Patterns 153
    11.2 Exploring Patterns with Contexts and Concrete Materials 153
    11.3 Connections between Geometry, Patterns, and Algebra 155
    11.4 Patterns and Number Properties 155
    11.5 Repeating Patterns and Pattern Core 156
    11.6 Growing and Shrinking Patterns 157
    11.7 Making Predictions about Linear Patterns 158
    11.8 Graphing Linear Patterns 160
    11.9 Pattern Rules 162
    11.10 The Variable 164
    11.11 Equation Concept 165
    11.12 Solving Equations with Manipulatives 166

Chapter 12: Algebraic Concepts 171
    12.1 Solving Equations Algebraically 171
    12.2 Ratio and Proportion 172
    12.3 Building Patterns That Change in Two Dimensions 176
    12.4 Describing Patterns That Change in Two Dimensions 177
    12.5 Graphing Patterns That Change in Two Dimensions 178
    12.6 Rates of Change, Slope, and First Differences 181
    12.7 Second-Degree Pattern Rules 184
    12.8 From Algebra Tiles to Algebraic Methods 188
    12.9 Exponents as Multi-Dimensional Numbers 191
    12.10 Exponentials 192

Chapter 13: Geometry 195
    13.1 Terminology 195
    13.2 Straight Lines and Angles 195
    13.3 Two-Dimensional Shapes 197
    13.4 Three-Dimensional Shapes 200
    13.5 Similarity and Equivalence 201
    13.6 Transformations and Symmetry 203
    13.7 Geometric Reasoning and Proof 206
    13.8 Technology and Dynamic Proof 207
    13.9 The Pythagorean Relationship 208
    13.10 Geometry in Art, Design, and Entertainment 210

Chapter 14: Measurement 214
    14.1 Linear Measure 214
    14.2 Rectangular Area: Exploring Covering and Counting 218
    14.3 Circle Area 223
    14.4 Relationships between Area and Perimeter 227
    14.5 Surface Area 231
    14.6 Volume 232
    14.7 Mass and Capacity 235

Chapter 15: Data Management and Probability 237
    15.1 Introduction to Data Management 237
    15.2 Discrete and Continuous Data 238
    15.3 Sampling and Representing Data 239
    15.4 Measures of Central Tendency 243
    15.5 Concrete Models of Probability: The Draw Ticket 245
    15.6 More than One: Dice Rolls 246
    15.7 How Many Ways? Full Circle Back to the
           Number System 248

References 252
Glossary 256

Preface

Mathematical Models for Teaching is designed for pre-service and in-service elementary teachers. The aim of this text is to support the development of a thorough understanding of the mathematics concepts required for teaching, with a focus on topics and concepts that research has shown to be most fundamental—and often problematic. It was our intention to produce a book that substantially increases pre-service and in-service teachers' comfort with the content they teach by helping them to develop a deeper, and more conceptual and connected understanding of elementary mathematics.

Mathematics for teaching is a specialized domain of mathematics that employs mathematical models and approaches that are distinct from, or may even conflict with, more general post-secondary mathematics teaching methods. Mathematics for teaching focuses particularly on using classroom manipulatives, models, and reasoning in mathematics that support students' learning. This text is for education and mathematics faculty who wish to better understand teachers' needs while instructing specialized courses in mathematics for teachers.

According to the National Mathematics Advisory Panel, the mathematical preparation of elementary teachers must include "ample opportunities to learn mathematics for teaching" (2008, p. xxi). A recent US report from the National Council on Teacher Quality on teacher preparation in mathematics states that teachers "must acquire a deep conceptual knowledge of mathematics … moving well beyond mere procedural understanding" (2008, p. 11). In Canada, the *Policy Statement for Canadian Elementary (K–8) Teacher Mathematics Content Development* suggests that

> knowledge of how mathematical understanding may develop in children, as well as of the models, representations and practices that support students' mathematical development are essential…. Appreciating and responding to alternative student approaches requires deep and flexible conceptual understanding on the part of teachers, as well as the ability to unpack students' thinking in order to recognize generalizable strategies or identify student misconceptions. (Kajander and Jarvis, 2009, p. 14)

Helping teachers develop such an understanding is the goal of this text. This book could be used as the main resource for a specialized mathematics course for teachers, as a supplemental resource for a curriculum and instruction course (along with a more standard methods course text), or to support in-service teachers' professional development.

While we have not attempted to address every aspect of elementary mathematics, we have included and deconstructed typically demanding content according to research on teachers' needs as identified by the National Mathematics Advisory Panel in their 2008 publication, *Foundations for Success*. We aim to promote reasoning and sense-making, rather than memorization, as the starting point for learning. Based on fundamental changes to classroom mathematics learning (National Council of Teachers of Mathematics [NCTM], 2000), it is our goal to equip teachers with the models and understanding necessary to facilitate students' construction of knowledge in a social-constructivist paradigm, as described by English, Fox, and Watters (2005) and Palincsar (1998).

The new mathematics methods described by the National Council of Teachers of Mathematics (2000)—variously termed *constructivist, inquiry, problem-based,* and *reform-based*—require teachers to have a much deeper and more connected conceptual knowledge of content. Teachers must be able to assess student-developed methods and unpack the mathematical elements of student thinking, including misconceptions, to successfully move students' learning forward. Developing a deep understanding of content is also the best way to address the anxiety that is sometimes associated with mathematics.

Moving beyond earlier conceptions of constructivism as a paradigm of mathematics learning, Lesh and Doerr (2003) describe a models and modelling approach, which provides the framework for this text. A large body of research further suggests that such a focus on conceptually oriented teaching via models and reasoning promotes skill development at a level equal to or greater than what is achieved when students receive only procedurally oriented instruction, as reported by Hiebert and Grouws (2007); hence, such "concepts, operations, and relations" (Hiebert and Grouws, 2007, p. 391) underpin this text. Our title, *Mathematical Models for Teaching: Reasoning without Memorization*, underscores the idea that skills and proficiency will result from a focus on models and reasoning, and should not form the initial goal or approach used in learning.

Pre-service teachers who have used earlier drafts of this text have consistently commented that it helped them see mathematics "in a different way." They noted that they learned that "there isn't just one way to grasp a concept." It helped them understand "how concepts connect," and to see things "that were never made explicit before."

A comprehensive chapter list encourages readers to work through the text in a sequential manner; however, this is not necessarily required. When teaching mathematics, it is not only important to focus on the content, but also the relational links and connections within and between concepts. We envision readers starting by examining particular topics, but then easily threading their way through conceptually related ideas as needed. Activities that demonstrate how to use content to construct appropriate classroom tasks are offered at the end of each section. We have used an informal, conversational writing style that we hope will inspire inquiry-based learning.

While this book is intended as a *mathematics* book, it draws on recommendations regarding the content, design, and delivery of effective mathematics courses for future elementary teachers (such as those provided by Hill and Ball, 2004; Kajander and Jarvis, 2009; and the National Council on Teacher Quality [NCTQ], 2008), so at times the style of the exposition may resemble that of a curriculum and instruction course for teachers. However, we emphasize that not all of the content that might be typically be included in a curriculum and instruction course is found here—we focus on the mathematical content elements of teacher preparation, rather than on more general teacher preparation and pedagogy. We anticipate that this approach will support the specialized nature of teachers' mathematical learning needs.

Acknowledgements

Over 700 teachers, both pre-service and in-service, participated in the research that supports this book, and we are grateful for their trust. In particular, the teachers in the whimsically named focus group the Association of Amazing Math Teachers, most notably Maria Casasola, Eric Fredrickson, and Elizabeth Petrone, were very helpful in informing the project. Funding from the Natural Sciences and Engineering Research Council of Canada through the University of Manitoba CRYSTAL grant is gratefully acknowledged, along with the support of the project's principal investigator, Gordon Robinson, and many associated colleagues, particularly Ralph Mason.

A number of other mathematics education colleagues, as well as mathematicians, also provided much-needed advice and support. Susan Oesterle provided detailed feedback and suggestions for most chapters, which were incredibly helpful. Peter Taylor and Miroslav Lovric patiently answered our mathematics questions, and Donna Kotsopoulos was always positive and encouraging. We are grateful to Tim Sibbald for reading an early draft, and to Elaine Simmt for her helpful comments.

This project could not have been completed on time without the editorial assistance of colleague and friend Jennifer Holm, to whom we are most grateful. We would also both like to thank our families for their encouragement and patience during the writing process. Tom is grateful to Todd, whose love and support are so often the wind beneath his wings. Ann is grateful to Wally and their friends in Mulberry for their patience and kindness while she was working.

Chapter 1

Introduction to Mathematics for Teaching

What Is Mathematics?

Teachers' beliefs about what mathematics is and how it should be learned are formed during their own schooling, and often form the basis of their own teaching practice. We have met many elementary teachers who tell us that teaching mathematics is not their strength. While they may be comfortable designing creative, interesting, student-based lessons in other subject areas, they say that they often feel less comfortable with their ability to do this within mathematics. Some other teachers feel that as long as they know "the rules," mathematics is not that difficult. In fact, many teachers were previously taught that mathematics is a rule-based subject, and that there is only one correct answer to any given problem and one method of deriving that answer. As a result, when individuals begin teacher preparation programs, they may believe that mathematics is about following procedures memorized by rote (Ball, 1988a, 1988b; Handal, 2003; Holm and Kajander, 2011, 2012; Thompson, 1992). Nothing is farther from the truth about mathematics! Mathematics is a rich and creative problem-solving-based subject. For descriptions of problem- and inquiry-based mathematics learning, see Ambrose, 2004; Askey, 1999; Ball and Bass, 2000; Boaler and Humphreys, 2005; Kajander et al., 2010; and Weiss and Moore-Russo, 2012.

What Students Need

During the past 25 years, research on the ways students understand and use mathematics (NCTM, 2000; National Mathematics Advisory Panel [NMAP], 2008) suggests that what they need to know about fundamental mathematics has changed and developed greatly. More than ever, students need to be good problem-solvers rather than "mini-calculators." They must be able to effectively represent a mathematics problem, often taken from a real-world context, to reason it out. They must be able to make connections among mathematical ideas in order to devise appropriate problem-solving strategies. They must be able to interpret and ultimately communicate their solutions. Such mathematical work not only requires students to master new and different types of mathematical understandings and approaches, but teachers must also do the same. This places significant demand on teachers, as described by Sowder (2007). Helping teachers to develop new ways of understanding based on inquiry and sense-making is the purpose of this book.

How Mathematics Is Learned

The way that practitioners use mathematics does not reflect the way it is initially developed and understood by learners (Skemp, 1986), regardless of whether the learners are schoolchildren or mathematicians (Weiss and Moore-Russo, 2012). Skemp illustrates this principle using the following example: We don't learn the concept of "colour" by reading a definition of light wavelengths; rather, we first need to experience and compare things that are red, yellow, blue, and so on before we can construct—and understand—the relatively abstract concept of colour.

The problem for teachers is that once we have learned a certain generalization or "rule," it seems simpler to use (and, therefore, teach) that rule than to provide a lot of problems for students to explore. Problem-based learning can feel messy and, at times, incomplete (Zack and Reid, 2003; Bay-Williams and Meyer, 2005). This may increase the appeal of teaching via procedures, because we, as teachers, might think this approach is easier; however, the critical point is that generalizations are only easier to understand and use after we have gone through the process of constructing their meaning for ourselves. Our minds are uniquely wired to make mathematical generalizations or abstractions, and to use them once we have done so—but this is not how these ideas are learned in the first place! Unpacking the higher-order ideas we have about mathematics to develop lessons that allow students to establish the conceptual building blocks themselves remains the central challenge of good mathematics teaching.

Teachers' Knowledge

Research suggests that teachers' knowledge must be broad in scope and include an understanding of how ideas are initially developed (Lesh and Doerr, 2003; Ma, 1999; Weiss and Moore-Russo, 2012). In our work, we have formed various ideas of what such knowledge should be, a central theme of which is deep and connected understanding, based on models and reasoning and applicable in the classroom. Our own ongoing research about best practices in supporting teacher mathematics development (Kajander, 2010a, 2010b; Hart, Swars, Oesterle, and Kajander, 2012; Kajander, Fredrickson, Casasola, and Boland, 2013; Kajander and Holm, 2013) illustrates the challenges faced by teachers, and provides grounding for this book. In particular, teachers need to know a multiplicity of models that are typically constructed by students, as described by Doerr and Lesh (2003), and how these apply to both mathematics and learning. Knowledge of such models is an aspect of the "distinct body of knowledge associated with mathematics teaching" (Davis and Simmt, 2006, p. 294). Kahan, Cooper, and Bethea (2003) explain how this new field of "mathematics for teaching" described by Ball, Thames, and Phelps (2008) differs from the knowledge typically taught in more general university mathematics classes. Hill, Sleep, Lewis, and Ball (2007) describe this new approach as including knowledge of the various ways students might understand a concept. Such a distinct body of knowledge forms the basis of this book, which is designed to support elementary teachers' deep and conceptual mathematical learning as needed for effective classroom teaching.

Supporting Research

This book is based on two major premises derived from substantial mathematics education research during the last several decades, one of which is about the nature of children's mathematics learning, and the second of which is related to the demands that such learning environments place on teachers' knowledge. The first body of research is about effective classroom learning. The US-based National Council of Teachers of Mathematics (NCTM) produced the first major publication related to mathematics reform in 1989, *Curriculum and Evaluation Standards for School Mathematics*, as well as the more recent *Principles and Standards for School Mathematics* in 2000. These documents were the culmination of several decades of research about children's learning across North America, and describe classroom learning that is far richer, more conceptual, more problem- and context-based, and that makes a greater number of connections between and among concepts than in the past. Further research, described by Lesh and Doerr (2003), supports the effectiveness of such classroom learning in the modern era, particularly for learners that have traditionally been less successful in mathematics. For example, students who were allowed to learn using inquiry-based methods with models and manipulatives were able to construct meaning for the ideas more effectively than when they were told rules. Curriculum documents all over North America have been revised and developed over the last several decades based on this research. These curriculum documents, which include a much broader and more conceptual approach to mathematics, as well as explicit emphasis on problem solving and *mathematical* models and processes, place significant demands on teachers in terms of the need for richer and deeper mathematics understanding. Kajander and Holm (2013) and Sowder (2007) illustrate the challenges teachers may experience in this regard, particularly if they were taught mathematics using only a rule-based approach.

In a reform-based learning environment, teachers are asked to help students develop mathematical ideas by means of a conceptually based problem-solving paradigm. While providing detailed information on planning and facilitating actual classroom mathematics lessons is not the purpose of this book (the further reading section at the end of this chapter contains books that provide more information about lesson construction), the mathematical understanding needed by teachers to construct and facilitate such classroom lessons is explored in depth.

The second premise of this book is that there is a body of specialized mathematical knowledge that teachers require to facilitate effective classroom teaching. In their research, Ball, Thames, and Phelps (2008), Davis and Simmt (2006), McNeal and Simon (2000), and Silverman and Thompson (2008) describe the development of the field of "mathematics for teaching." Ball, Thames, and Phelps define various types of knowledge for teachers, including a specialized body of content knowledge unique to teaching, which "is mathematical knowledge not typically needed for purposes other than teaching" (2008, p. 400). They further state that teachers' mathematical work "involves a kind of unpacking of mathematics that is not needed—or even desirable—in settings other than teaching" (2008, p. 400). Hence, while the work of a research mathematician might involve developing increasingly powerful and abstract generalizations, the work of teachers involves "prying apart constructs, [and] making sense of the analogies, metaphors, images, and logical constructs that give shape to a mathematical construct" (Davis and Simmt, 2006, p. 301). This knowledge is indeed mathematical; Ball, Thames, and Phelps (2008) claim that "deciding whether a method or procedure would work in general[,] ... determining the validity of a mathematical argument, or selecting a mathematically appropriate representation ... requires mathematical knowledge and skill, not knowledge of students or teaching" (p. 398). Such specialized mathematical knowledge helps teachers construct good classroom problems while anticipating students' various potential interpretations, and recognizing what students must already understand in order to take on these problems. Teachers also draw on their content knowledge to effectively help students model and reason through problems, recognize their errors, and provide prompts, questions, or further or different problems that will allow them to recognize and clarify their misconceptions. Teachers support effective classroom discussions about possible outcomes and solution methods of the problems, and ultimately assess student progress in order to plan appropriate new problems. Research is becoming more and more conclusive, gathering evidence that suggests that such deep and particular teacher mathematics knowledge supports measurably stronger student achievement in this subject (Baumert et al., 2010; NMAP, 2008; Sowder, 2007).

Determining precisely what content to include in this book has been the focus of a number of our recent research projects. For the last 10 years, we have been tracking future elementary teachers' understanding of mathematics as these prospective teachers enter teacher education programs. We have found, over and over, that many teacher candidates initially struggle to explain the concepts in Ontario's elementary curriculum, and similar results are found elsewhere, as illustrated by the research of Ma (1999) and Weller, Arnon, and Dubinsky (2009). We have worked in parallel with elementary classroom teachers who we judged to be highly effective teachers, chronicling what they know, and how they use this knowledge when teaching mathematics in their day-to-day classroom work with students. This work, along with recommendations from documents such as *Foundations for Success*, the report of the National Mathematics Advisory Panel (2008), helped us to decide what to include in this book. Supporting the development of a specialized mathematical understanding, conducive to effective classroom learning, is the primary focus of this book.

Concepts and Procedures

Ball and Wilson (2012) have explained that in mathematics education, possibly more than in any other subject, there is a significant gap between what we know from research about best practices and what current classroom practice generally looks like. Research in mathematics education over the past three decades indicates that most students learn best (and retain ideas longer) if they are able to construct deep understanding of the concepts themselves, rather than simply memorizing methods and rules. When students learn conceptually, they actually develop the procedural skills along the way. Procedural fluency has been shown to develop alongside conceptual understanding even when it is not the focus of instruction (Stein, Remillard, and Smith, 2007). Students retain computational knowledge and skills developed in this manner longer, and can apply the procedures appropriately and with greater accuracy when solving new problems or those that are not routine problems.

Developing a deep understanding of the fundamental concepts may at first seem like a daunting challenge for some teachers, especially those with "math anxiety," as teachers who are anxious about their mathematical capacity may have survived their own mathematical education largely by memorizing. Also, it may at first seem somewhat counterintuitive to go "backwards" (Ball and Wilson, 2012)—backing up and unlearning or letting go of a previously memorized method—when trying to figure something out that we thought we already understood. Teachers may initially experience an internal resistance when working to construct deep understanding of previously memorized mathematical ideas (Holm and Kajander, 2011, 2012); Eisenhart and colleagues (1993) describe how it can be challenging for teachers to focus their own instruction on these ideas. For example, it seems simpler to accept the rule that states that "the product of two negative integers is positive" than to attempt to develop a model to explain the reasons behind this rule; however, the mathematical power gained from deeply understanding a concept soon makes the effort worthwhile.

Mathematics Anxiety and Mathematical Development for Teachers

Mathematics anxiety may stem from an awareness of inadequate personal understanding of mathematics, and the best way to address such anxiety is by developing enhanced content knowledge (Sowder, 2007). According to Stipek, Givvin, Salmon, and MacGyvers (2001), understanding *how* previously memorized rules work contributes to the development of more effective teaching practices. In our experience nothing addresses the "math anxiety" experienced by some teachers more effectively than developing, often for the first time, a conceptually deep, connected, and flexible understanding of previously memorized rules. See Holm and Kajander (2012) for a detailed description of pre-service teachers' experiences as they navigated such a paradigm shift, and how emotionally rewarding it was for them.

Teachers have the responsibility to understand mathematics in a way that will allow them to connect the mathematical ideas to students' levels of development, as well as support the conceptual development of ideas. The process of backing up and redeveloping the models, concepts, and connections in mathematics is not trivial. In some ways, the more sophisticated one's procedural mathematical fluency is, the harder and more counterintuitive this unlearning process may initially seem. Moreira and David (2008) suggest that a strong procedural knowledge base may actually inhibit the desire to unpack and redevelop the underlying ideas, and further argue that strong procedural skill may even block a prospective teacher's willingness to engage with the mathematics more conceptually. However, it must be remembered that the construction of deep and connected mathematical knowledge, as described by Ma (1999), by students and teachers alike, is extremely powerful with respect to both teaching and learning. For students, making sense of mathematics in this way (instead of blindly memorizing rules) supports learning that is rewarding and well retained, rather than frustrating and easily forgotten. We can only assure the reader that engaging in this process will increase both personal understanding of mathematics and the capacity to provide better mathematical learning opportunities for students.

Reasoning and Rigor

One criticism of problem-based learning as the new paradigm for learning mathematics more conceptually is a perceived lack of mathematical rigor and the ability to produce complete and syntactically correct arguments according to generally accepted standards. Traditionalists may argue that rigor is lacking in problem-based learning environments. Historically, higher-level mathematics was taught by exposing students to relatively sophisticated mathematical arguments, such as proofs; this often consisted mainly of examining those proofs already in existence (Beaugris, 2013). This method was thought to teach mathematical reasoning, and instill in students a sense of what achieving mathematical rigor entailed.

A flaw in this traditional approach is a lack of understanding of human mathematical development. While examining a sophisticated mathematical argument or proof presented in formal language might be interesting and illustrative, it does little to help students develop the ability to construct valid mathematical reasoning and rigorous arguments (Skemp, 1986; Reid, 2011). To use a metaphor, while budding painters might benefit from viewing famous works of art at a gallery, such viewing is hardly a complete educational experience for an artist. The same is true in mathematics. Children need to have the very experiences mathematicians themselves enjoyed when they constructed well-known and commonly used mathematics ideas, methods, and arguments. They need to experiment, conjecture, model, test, look for exceptions, find patterns, construct and extend arguments, and generalize—to name but a few mathematical processes highlighted by the NCTM in their 2000 publication. Ultimately, such skills are the very ones that will allow students to construct mathematical arguments of their own. Such mathematical arguments, based on connected understanding as well as sound reasoning, are the very underpinnings of mathematical reasoning and proof.

Student-generated arguments can become more and more rigorous as students' mathematical sophistication increases (Zack and Reid, 2004), just as, historically, mathematics became more rigorous as the field progressed. Research suggests that such a progression is an effective strategy for learning about the process of mathematical proof (Stylianides and Ball, 2008; Reid, 2011).

In this book, we often employ the stepping stone of a *generic example* (Reid, 2011) to argue an idea using a specific instance, while simultaneously thinking about how the idea might apply more generally. A generic example is a specific example constructed in such a way that it is representative of many other examples, such as a triangle with no special properties (e.g., that is not equilateral or right-angled).

> **MATHEMATICAL TERM**
>
> A *conjecture* is an idea or theory proposed as a mathematical truth.

Summary

To construct classroom learning environments that support deep and connected learning for students, teachers must have enhanced mathematical understanding. In their daily work, teachers need to be able to design useful contexts and examples, provide models and strategies, support alternate student methods and solutions, pose good questions, and help students identify misconceptions. Such work requires a deep, rich, and specialized understanding of mathematics. Teacher knowledge of mathematical concepts, models, representations, strategies, and connections directly supports student learning, and is therefore critically important to student success. Providing teachers with the means to develop this enhanced knowledge of mathematics is the purpose of this book. This book will also help all learners of mathematics think about elementary mathematics in a more meaningful way—reliance on a good memory is not required!

Further Reading

Ball, D. L., Hill, H. C., and Bass, H. (2005). Knowing mathematics for teaching: Who knows mathematics well enough to teach third grade, and how can we decide? *American Educator, 29*(3), 14–46. http://www.aft.org/pdfs/americaneducator/fall2005/BallF05.pdf

> This article describes and provides examples of what is particular to teachers' specialized knowledge of mathematics for teaching.

Holm, J., and Kajander, A. (2012). Interconnections of knowledge and beliefs in teaching mathematics. *Canadian Journal of Science, Mathematics, and Technology Education, 12*(1), 7–21.

> This article describes the journeys of selected pre-service teachers as they strengthen their content knowledge base during teacher preparation.

Small, M. (2012). *Making math meaningful to Canadian students, K–8.* 2nd ed. Toronto: Nelson Education.

> This comprehensive textbook would be useful as a companion resource to *Mathematical Models for Teaching*. It provides information about classroom mathematics teaching strategies.

Chapter 2
Introduction to Mathematical Reasoning

2.1 Mathematical Models

An important part of a mathematician's tool kit is the *mathematical model*, which can support thinking and reasoning (Doerr and Lesh, 2003). A model can be a diagram, sketch, graph, or any other visual aid. It can also be a mental image, an abstract construct, or a computer simulation. A model provides a way to think about and work with a problem. Early use of a concrete model or manipulative can contribute significantly to the development of conceptual understanding of a more algebraic model, such as a set of equations, in the longer term. Constructing a concrete model or diagram is often the aspect of problem solving that requires the most creative thought or inspiration. Creating a model supports the development of a mental construct for a problem and its solution.

Mathematical thinking can occur very rapidly or over a sustained period of time. Mathematicians often physically move around while they are thinking out new ideas, and discuss aspects of these new ideas with colleagues. Discussions can be lengthy, or can happen intermittently. When mathematicians discuss ideas, they do not always use chalkboards or whiteboards, but often rely on quickly made diagrams (models), hand gestures, and other forms of communication. In other words, the process of mathematical thinking is collaborative, interactive, and often involves movement and visual aids. Like mathematicians, students also need active learning environments and supportive working conditions to effectively problem solve, and thereby construct knowledge.

Creating and using models effectively is an important aspect of mathematical problem solving. Although it may be possible for students to imitate solution methods, basically memorizing them, without the use of models or other aids that support understanding, such an approach is often debilitating in the long run. In fact, memorizing procedures without understanding why these procedures work puts students at great risk of developing math anxiety or math phobia. In today's world of rapidly changing and ever-developing technology, we no longer need students to be mini-calculators; rather we need to help them to think, reason, and problem solve. Mathematical models are fundamental tools for doing so.

Example

The following is a very simple example of a problem solution using a "thinking and modeling approach" (Lesh and Doerr, 2003) to problem solving, as opposed to a merely "calculative approach." The problem might be called a *missing addends* problem:

Maria noticed that the new bag of 12 mini brownies was open and 7 brownies were left. How many brownies were eaten?

Students might ask if this is an addition or a subtraction problem. Using a model, it becomes clear that it can be either. Here are two possible student models of this problem:

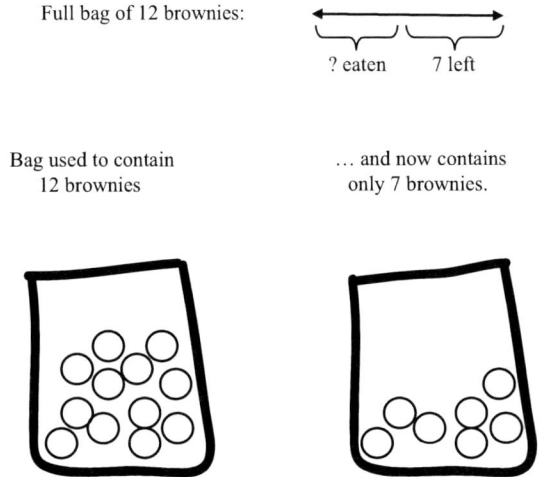

Follow-Up and Discussion

Either of the two sample models might lead students to develop a variety of numeric strategies. For example, a student might use the first model to start at 12 and count back to 7. Using the second model, a student might remove 7 brownies from the bag, and then count the remaining brownies. This removal of 7 might be aided by making a one-to-one correspondence with the 7 brownies on the right side of the diagram. A number of operational strategies can be used to solve the problem, based on the information presented in the models. The following are possible ways to calculate the answer to the problem:

$$7 + ? = 12 \text{ and } ? + 7 = 12$$
$$12 - 7 = ? \text{ and } 12 - ? = 7$$

It is important to note that the thinking and understanding aspect of the solution happens largely when the student creates a model, and that one student's model may be very different from another's. After constructing a model, students can use the related skill of translating what the model is showing into a numeric expression that actually calculates the answer to the problem.

Students who have difficulty with word problems tend to have greater difficulty with the modelling part of the process than the operational aspect. Once a model has been constructed, determining the final answer may not be that difficult. The use of models also makes it easier for students to communicate their thinking when they explain their approach to teachers and classmates.

In summary, focusing on the *modelling* aspect of mathematical thinking—rather than the calculating aspect—supports the development of problem-solving skills and deeper understanding. Students can use models to discuss and compare their thinking and strategies. Teachers sometimes explicitly suggest certain models and strategies for students to use, while at other times this choice might be left entirely to students. The emphasis on models highlights the mathematical *reasoning* aspect of student solutions, establishing thinking and problem-solving processes as the most critical aspects of learning.

2.2 Mathematical Communication

Another important aspect of mathematical learning is the communication of reasoning. When students are asked to explain their thoughts as they worked to solve the example provided above, initially they may simply say "First I took the 12, then I subtracted 7 …." In other words, they may describe their *calculations* rather than explaining their *reasoning*. Helping students refer to models as they explain their thinking supports the idea that mathematical reasoning, rather than procedure use, is the most important aspect of learning, and using the models while explaining may prompt students to communicate their reasoning in greater detail. Reasoning, justifying, and generalizing are the fundamentals of mathematical thought and learning, and are examples of mathematical *processes* (NCTM, 2000). Being able to describe and defend one's arguments is part of mathematical communication, which is another mathematical process that is also critical to development.

2.3 From Examples to Generalizations

There is the old and somewhat silly mathematical joke that since the odd numbers 3, 5, and 7 are all prime numbers, then all odd numbers must be prime. Of course, examining the next odd number, 9, immediately disproves this generalization. Finding one case in which an idea does not work—called a *counterexample* in mathematics—is enough to prove an idea wrong. But when can we be sure an idea is *right*—in other words, that it is always true? How much evidence is enough?

The key idea with respect to a mathematical generalization is that a concept or method must always apply in the specified context, regardless of what numbers are used. Ultimately, mathematicians construct more formally written mathematical arguments called *proofs* to verify new mathematical truths. But proofs are usually the final stage in ensuring the validity of new ideas; usually a great deal of exploration, modelling, and reasoning precedes these formally stated arguments.

To determine when enough cases have been examined to be reasonably sure that a concept always holds true, it is important to have the ability to choose appropriate examples. Examples that work well to illustrate that an idea is more generally applicable are sometimes called *generic* examples (Reid, 2011). Generic examples are numeric examples for which the reasoning used would also hold true more generally. The use of models that would work in general—that is, for any specific values—also helps in generalizing. Thus, a useful strategy to help students begin to reason mathematically, even before they are developmentally ready for very formal mathematical arguments and proofs, is the development of the use of generic examples and models. In contrast, use of a non-generic example would be like trying to prove a property about triangles in general by using only an equilateral triangle. In selecting a generic example, it must be reasonably evident that the properties of that particular example *that are important to the argument* would still hold with other suitable instances of the idea. In other words, the reasoning about the example must be generalizable.

Take a moment to sketch a few different parallelograms—four-sided shapes with opposite sides parallel.

After looking at the examples you have drawn, it seems reasonable to say that no matter the size and shape of a parallelogram, it can *always* be cut into two triangles. It is reasonable to generalize then, that this will always be possible. We might thus informally generalize that every parallelogram, no matter the size and shape, can be divided into two identical triangles.

> **MATHEMATICAL TERM**
>
> A *parallelogram* is a four-sided shape with both sets of opposite sides equal and parallel. A parallelogram may or may not be rectangular (having 90-degree or right-angled corners).

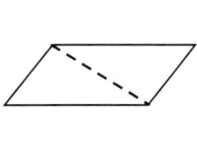 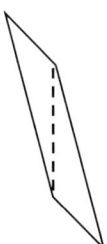 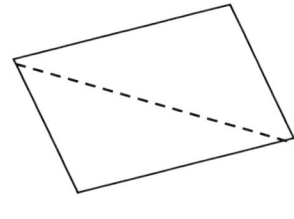 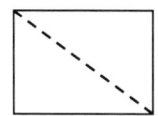

Our example shows that a parallelogram has four sides and thus four vertices. Hence, by joining two opposite vertices with straight lines, we will have constructed two triangles. Take a moment to explore this idea by drawing a few more different parallelograms. In each, draw a line to join the opposite vertices, noting that two triangles are formed. Think about how you might reason that two triangles will always be formed by joining opposite vertices in a parallelogram. Of the parallelograms that you have drawn, choose one that does not have any special properties, such as a rectangular or square parallelogram. Such a general picture of a parallelogram, for which the reasoning about the construction of two triangles by joining opposite vertices holds, and which also implies that the reasoning will *always* hold true in a similar case, is an example of a generic diagram or model.

The importance of mathematical generalizations is that they allow us to know that something will *always* work or hold true. This is the power of mathematics. On occasion, however, even the most seemingly sound ideas in mathematics prove to be untrue. Thus, to be sure, mathematicians must *prove* things beyond all doubt. The abilities needed to ultimately construct the reasoning behind formal mathematical proofs are the very abilities that students develop during problem-based classroom learning experiences.

2.4 Argument, Proof, and Rigor

The constitution of formal mathematical proofs, and the accepted style and allowable arguments in proofs, have evolved over the centuries. Hence, there is a certain cultural element to what is considered an acceptable proof. It is generally accepted that a formal mathematical proof consists of a set of assumptions (either previously proven ideas, called theorems, or what are considered acceptable ideas, called axioms), followed by reasoning using accepted rules of mathematical argument, leading to an inevitable conclusion or result, which then becomes a new theorem (Beaugris, 2013).

Newer forms of proofs are also emerging. For example, there is an increasing tendency to use computer software to help students—and mathematicians—model, explore, and reason mathematically. Software, such as a program called Geometer's Sketchpad, is available that allows students to explore geometric concepts by constructing and manipulating shapes. Because this software allows students to move, manipulate, and transform images, it is called *dynamic software*. In geometry, the concept of a *dynamic proof*—the use of a diagram that we can move randomly, but that still illustrates a given property—is becoming more accepted as a valid form of mathematical reasoning and argument. For example, if we draw a parallelogram and join the opposite vertices with a line, we observe that no matter how we change the parallelogram in size or location, it will still contain two triangles. Hence, we can use this dynamic diagram of a parallelogram divided into two triangles to reason, and thus dynamically prove that a parallelogram will always contain two triangles.

The key element of a mathematical proof is an argument that can be applied *in general*. This key aspect of generalizing is an important part of the process of reasoning and proving that can be effectively incorporated into students' classroom work. Zack and Reid (2003, 2004) found that students also wanted their proofs to explain a concept and why it worked. As students ask questions and work to understand one another's ideas by describing, justifying, and defending their ideas to one another, a true mathematical culture emerges that is based on investigating and generalizing using accepted methods of mathematical reasoning. Thus a problem-based or inquiry-based mathematics classroom is exactly the type of environment that, rather than impeding the learning of rigor, actually *best supports* the development of mathematical argument, rigor, and, ultimately, an understanding of proof.

Chapter Problems

1. A child reasons that the answer to 25 − 18 is 2 + 5 or 7. Draw a model that the child may have used to help with such reasoning.
2. Choose a mathematical calculation method with which you are familiar, for example, the procedure for multiplying two numbers between 11 and 99, and think about how you could provide reasoning to show how and why this method always works.
3. Draw a 10-by-10 grid. Explore how many squares of any size could be drawn in the grid. (You don't have to be able to draw them all at once). For example, you could draw 1 square that is 10 by 10 (the entire grid), and 100 squares that are 1 by 1. Look for a pattern, and explore whether the pattern would hold for grids of different sizes.

Further Reading

Bennett, C. A. (2012). Using tiered explorations to promote reasoning. *Mathematics Teaching in the Middle School*, *18*(3), 166–173.

> Bennett discusses how he used a "crime scene for mathematics investigation" to both differentiate his mathematics classroom and improve student reasoning in mathematics. He gives examples of lessons, as well as practical procedures used in the classroom to support all students in learning mathematics.

Rathouz, M. (2011). 3 ways that promote student reasoning. *Teaching Children Mathematics*, *18*(3), 182–189.

> Using concrete examples, Rathouz discusses three strategies that can be used to encourage student reasoning within the mathematics classroom.

Rwading, M. R., and Wills, T. (2012). Discourse: Simple moves that work. *Mathematics Teaching in the Middle School*, *18*(1), 46–51.

> Rwading and Wills discuss an example classroom where students excel at conversing in mathematics to set up the basis for their article. The article focuses on strategies for developing a community in which students can use discourse effectively in their mathematics classrooms.

Small, M. (2012). Visual reasoning K–12. *Ontario Mathematics Gazette*, *51*(1), 21–24.

> Small uses examples of students at different levels in K–12 classrooms to discuss reasoning in mathematics, and demonstrates the necessity of visual reasoning at all levels of mathematics.

Stoner, M. A., Stuby, K. T., and Szczepanski, S. (2013). The engineering process in construction and design. *Mathematics Teaching in the Middle School*, *18*(6), 332–338.

> Stoner, Stuby, and Szczepanski detail a Grade 8 project that used mathematical thinking processes in hands-on engineering-related contexts to increase student engagement and broaden mathematics concepts within the classroom.

Zack, V., and Reid, D. A. (2003). Good-enough understanding: Theorising about the learning of complex ideas (Part 1). *For the Learning of Mathematics*, *23*(3), 43–50.

> Zack and Reid provide examples of students' reasoning when exploring challenging mathematics problems.

Chapter 3
Introduction to Numbers

3.1 From Quantity to Number Symbols

Quantity determines how many or how much. One way to keep track of a quantity is to draw pictures, a method that was used in ancient times. Since this quickly becomes cumbersome, different types of symbol systems that allow us to keep track of numeric amounts more efficiently have been introduced over the ages. Our own number system, called the decimal system, or base ten, is but one way to do so. In the decimal system, we use the symbols 1, 2, 3, 4, 5, 6, 7, 8, 9, and also 0, which was introduced later. The idea of the decimal system is that to record more than 9, we start counting again from 0 in the *next* column to the left; so, the next number after 9 is written 10. The decimal system is likely based on the 10 fingers of human hands. Base ten is not the only possible number system. If we had only 6 fingers, we might have developed a counting system that looks like this: 0, 1, 2, 3, 4, 5, 10, 11, 12, 13, 14, 15, 20, ... 54, 55, 100, 101, 102, 103, 104, 105, 110, and so on. In this system, the numeral 10 would mean the next number after 5, or what we mean by the symbol 6 in base ten.

While the idea of using symbolism to represent quantities may well be mathematically universal, the choice of a *particular* number system to represent quantity is unique to a specific human history and culture. The use of a symbol, such as a number, to represent an amount is one of the first mathematical ideas (or *abstractions*) a student might encounter. The decimal system is by no means the only number system possible; asking students to investigate other number systems is an excellent student enrichment idea. But the decimal system, the universally adopted number system in North America and Europe, forms the basis of this book.

3.2 Quantity, Digits, and Place Value

Basic perceptions of number and quantity—as well as some representations of these—are innate. For example, it has been shown that babies are naturally able to distinguish one dot from two dots—although, of course, they don't yet know the symbols 1 or 2. Our basic counting numbers, 1, 2, 3, and so on, are symbols that help us keep track of complete or whole amounts. This group of numbers, starting at one and going on and on, is called the *natural numbers*. Sometimes mathematicians call it the *set of natural numbers*. When zero is added, this set or group is called the *whole numbers*.

The *digits* available in the decimal system are the symbols 0, 1, 2, 3, 4, 5, 6, 7, 8, 9. If we have, for example, three of these digits in a number, such as the number 275, we call it a three-digit number. There are assumptions made here about the use of the digits; this notation has been *constructed* over time to have a particular meaning. This particular interpretation or meaning is called *place value*.

Place value refers to meaning that results from the combination of the *value* of a digit and its *location* in the number. For example, the number 13 does not, of course, mean a 1 and a 3, as some students may initially think. It is important to realize that while initial ideas of quantity, such as the idea of 1 or 2, may be innate, the construction of the number system based the concept of place value is not. In the number 13, the 3 indicates three units or things, but the symbol 1 does not mean simply one; its location in the number 13 means one group of 10. Another way to think of the number 13 is that it means 10 + 3. Using similar reasoning, the number 275 means 5 (or 5 ones), plus 7 groups of ten, plus 2 groups of ten tens (and if we line up ten of the tens together, we can see that this is the same as one hundred). Common classroom manipulatives for modelling number values will be explored next.

3.3 Exploring Models and Manipulatives

Wooden or plastic models of quantities, such as unit cubes or centicubes (cubes that measure 1 centimetre by 1 centimetre by 1 centimetre) are often referred to as *concrete* materials or *manipulatives*. But it must be remembered that such materials are only *relatively* concrete. If we use three centicubes to imply three cookies, the cubes are still an abstract representation of the three cookies (and much less tasty!). The three centicubes help us make a mental bridge or connection between the quantity or number of cookies and the (relatively more abstract) numeric representation of 3.

The centicube is often used to represent the unit, or 1; 10 such cubes lined up represents 10. The 10 cubes together look like one long piece, hence, this is referred to as a *long*, or sometimes, a *rod*. Ten of the longs placed side by side look like a flat arrangement of 10-by-10 or 100 centicubes, and this hundreds piece is called a *flat*. Lastly, 10 flats (with each having 100 centicubes) can be stacked to form the thousands cube, which is sometimes called the *large cube* or *block*.

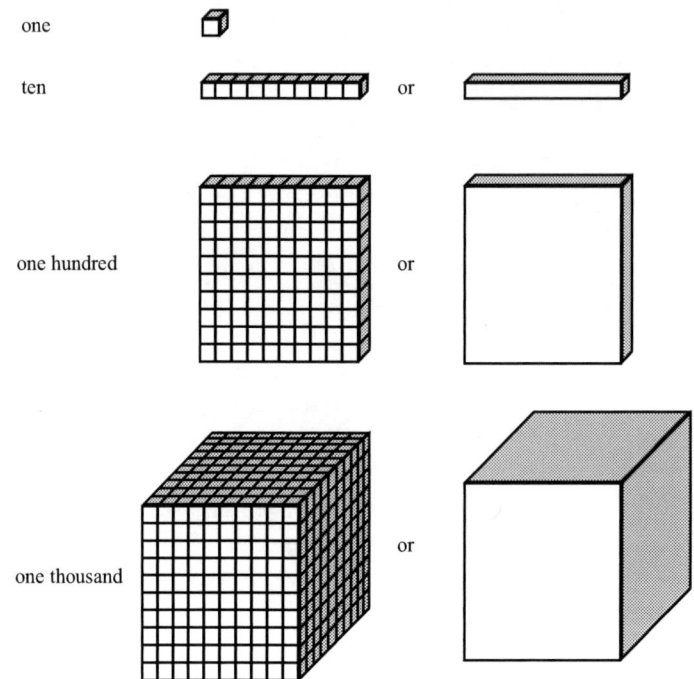

Below is a picture of 275, using a drawing of base ten blocks. The key idea to think about as you look at the picture is that 275 is composed of 5 ones, 7 groups of ten, and 2 groups of one hundred.

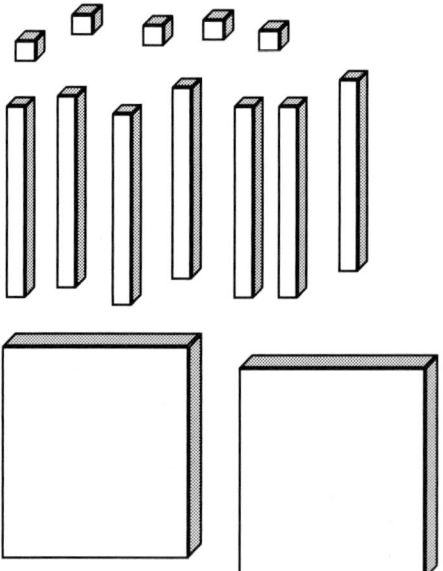

The purpose of using concrete materials in the mathematics classroom is to help students make bridges and connections between ideas that are relatively more and less abstract. For example, some students might be able to move from the three cookies concept immediately to the numeric representation 3. It may help to use three centicubes to illustrate this concept, as some students like to move and touch objects as they think. Others are better at visualizing, while still others might prefer to draw a picture on paper. The choice of whether or not to use a model or other materials—and which model or material to use—as an aid while thinking through and developing a mathematical idea is very personal, and highly dependent on a learner's thinking style and developmental level. While the teacher might make suggestions, the best aid should ultimately be determined by what is best for that learner.

Some teachers worry that students will keep using the materials when, according to a particular curriculum, they should be working symbolically. Teachers may ask, "When should I take the manipulatives away?" The answer is, "Never!" This is the beauty of mathematics learning; when students are ready to work more symbolically, they will spontaneously do so. No one wants to bother pushing cumbersome materials around when an elegant mathematical procedure is available. On the other hand, forcing students to use a procedure that they do not understand, when they still need concrete materials to think through the ideas, is detrimental in the long run. Procedural use *without* understanding is fragile and easily forgotten, and is highly likely to lead to errors.

In this book, we suggest ways to represent and model mathematical ideas and concepts for learners. Sometimes these models will be in the form of pictures and drawings, and at other times, we will suggest readily available concrete materials. The key point is that learners should be introduced to many ways to help them represent and develop their thinking, and encouraged to use those that make the most sense to them. When students are ready to think about a concept or method in a more general or abstract manner, they will spontaneously stop using the related models and materials.

3.4 Zero

The use of a symbol or digit to represent zero is more recent in mathematical history than the development of the other digits. Perhaps this is because in some ways the concept of zero is more abstract than the concept of one.

The use of the symbol 0 in number representations is tied to an understanding of place value. Students who use the digits of a number without regard to the location of the digits may not understand the difference between, for example, 37 and 307. The role of 0 in 307 is to imply 0 tens, so we don't get mixed up about the role of the 3 (which means 3 hundreds). As a number, 0 is sometimes an exception to the rules, as we will see in future chapters. As a result, the number 0 is a source of mathematical interest and curiosity.

3.5 Between Zero and One (or between One and Two, or between Two and Three, and So On)

Even very young children can understand the concept of half a cookie, or a pizza that is shared among 4 friends. If we have a very long piece of licorice, we could share it with 10 friends. If it was a metre long (100 centimetres) and we cut it into 1-centimetre pieces, then 100 children could have a taste. And if we cut each of those 1-centimetre pieces in half, then 200 children could have a tinier taste. How far can we go?

We can only cut concrete items, such as candy, into so many pieces before we are unable to divide it any further. But this is not so with mathematics. In mathematics, we can always cut something in half or find a number in between; to calculate half of any given number, we can divide it by 2. Of course, if we cut a quantity in half, we won't necessarily have a whole number anymore. We'll need some other ways to write the numbers in between the whole numbers. At this stage, the important idea is that there *could* be amounts between the whole numbers and that, mathematically, we can keep calculating them—forever! The numbers between whole numbers can be expressed as fractions or decimals. These ideas will be discussed further later in the book.

3.6 Infinity

To help students begin to think intuitively about the mathematical concept of infinity, use the following demonstration: Have a child stand between two large wall mirrors facing each other. The child will see his or her image reflected back and forth, and back and forth repeatedly—the image appears to go on and on. At a certain point, because the reflection is not perfect, it does get blurry and hard to see. Unlike the image in the mirror, in mathematics the idea of "on and on" or "forever and ever" is quite possible. Such is the notion of infinity. For example, we can have as many digits as we want in a number—although eventually it might be difficult to write down.

Defining things that are infinitely big or infinitely small has historically been a challenge in mathematics. There are certain ideas about infinity, in particular things that are infinitely close to each other, that seem contradictory or paradoxical. For example, an ant walks from point A to point B, but goes only halfway before stopping. It then travels half of the remaining distance, and stops again. Mathematically, if the ant continues to walk only half of the remaining distance, it will always have some tiny distance still to go. But practically, this seems to be a contradiction—surely the ant will eventually arrive at point B! More will be said about these ideas in chapter 10.

While the highly abstract idea of infinity has fascinated mathematicians for centuries, it is an interesting topic for students to mull over. They can discuss, for example, what it might mean to have numbers get bigger and bigger forever. Or they can be asked, "How close can two numbers get to each other without being equal?" Students' sense of these ideas will develop as they grow mathematically.

Chapter Problems

1. Choose a few whole numbers and practice representing them using base ten blocks. You might explore whether more than one representation is possible; for example, 14 can be shown as 1 ten and 4 ones, or as 14 ones.
2. Determine as many contexts as you can that would allow children to visualize or represent the idea of something happening an infinite number of times. One example is to have a student stand between two large mirrors, viewing his repeated reflection (as outlined in section 3.6).

3. Use 3 toothpicks to create a model of a triangle. Next to it, use 4 toothpicks to create a square, then create a regular pentagon (5-sided polygon shape) and a hexagon (6-sided shape). Build several more polygons with an increasing number of sides. As you work, look at how the pattern of shapes is changing. If you used 100 toothpicks, what would the resultant shape look like? What about 1,000? Think about the connection between a shape with an infinite number of "infinitely small" sides and a circle.

Further Reading

Bofferding, L., Kemmerle, M., and Murata, A. (2012). Making 10 my way. *Teaching Children Mathematics, 19*(3), 164–173.

Bofferding, Kemmerle, and Murata detail the experiences of three teachers who use activities to help kindergartners develop number concepts up to 10 using different addition and subtraction scenarios in context. The article includes the teachers' reflections as they discuss their classroom experiences.

Kajander, A. (2012). Math by and 4 teachers: Areas and volumes; The power of modelling. *The Ontario Mathematics Gazette. 51*(1), 18–21.

Using the context of area and volume, this article illustrates how some mathematics problems are much easier to solve using concrete models.

Gadanidis, G., and Gadanidis, M. (2011). *To infinity and beyond: The real story of Rapunzel.* Smart Math #10 (Kindle Edition). Brainy Day.

The authors illustrate the power of the imagination when conceptualizing mathematical ideas, such as the concept of infinity, in this engaging story.

Chapter 4
Whole Number Addition and Subtraction

4.1 Addition: Concept and Contexts

Addition is combining two or more quantities, or bringing two or more amounts together.

We can model addition by representing *all* of the quantities to be added—either concretely using manipulatives, with pictures and models, or symbolically—and then showing the amounts combined. Such modelling helps to illustrate that the quantities can be combined *in any order*.

When students use models to construct meaningful understanding of this big idea—the addition principle—number sense is enhanced. Students realize, for example, that contrary to how it is demonstrated in traditional teaching, it is not necessary to add a group of numbers from top to bottom or from left to right, and sometimes other ways are more efficient. We will discuss this in further detail in sections 4.2 and 4.3.

The concept of place value is an important companion to addition. For example, while $10 + 6$ is 16, $17 + 6$ is not 113. (Can you identify the misconception in the latter example? Thinking of the problem as $10 + 7 + 6$ may help clarify that the solution is $10 + 13$, and not $100 + 13$.) Playing games that involve the use of manipulatives and diagrams helps to reinforce this idea; one way to think of place value, for example, is that there is simply only space for 9 of whatever we are counting (for example, centicubes) in the ones column. After that we must gather them up in packages of 10 and place these packages in the tens column. The second column is called the tens column because it counts *groups of 10*. Since 1 more after 9 (or the next number after 9) no longer fits in the ones column, we either need a new symbol, or a new column. The largest amount we can record using only the first or ones column is 9, a one-digit number. After 9, since we have run out of symbols (at least in the decimal system), we go back to 0 in the ones column and record the 10 as a one in the next column. The value of the digit in the tens column tells us how many of these packages of 10 ones we have. Similarly, only 9 of these packages of 10 fit in the tens column; after that we must combine or group 10 packages of 10 and record this as a 1 in the hundreds column. And so on. So the 1 in 100, for example, means 100 ones, or 10 tens.

MATHEMATICAL TERM

A *sum* is the result of combining or adding quantities.

INSIGHT

Numbers can be added in any order.

If we were working in a base larger than base ten, say base sixteen, we would need enough one-digit symbols for another six values after 9. In other words, we would need a one-digit symbol instead of the two-digit 10 for the next number after 9. We might use the letter A. Then, instead of 11, we could use the symbol B, and instead of 12, we could use a C, and so on to 15. For the next number after 15, we can regroup to 10, which means, in base sixteen, one complete group of 16; so the base ten numbers 10, 11, 12, 13, 14, 15, 16 could be written as A, B, C, D, E, F, 10, in base sixteen. We need the extra symbols because we do not write a "1" in the second column until we have a complete group of 16. In base sixteen, the notation "10" really means one group of 16 and zero ones. Sometimes this is written 10_{16}, so we know it's base sixteen; the subscript 16 indicates that we are counting sixteens rather than the usual tens. The investigation of bases other than 10 is a wonderful enrichment topic; for example, students will find it interesting to learn that computers use base two. This means they can only use two symbols, 0 and 1, and must regroup after 1. For demonstration purposes, here are the first 10 values in base two: 1, 10, 11, 100, 101, 110, 111, 1000, 1001, 1010. The last number, for example, means $1 \times 2^3 + 0 + 1 \times 2^1 + 0$.

Addition is all around us in our everyday lives. Students can quite naturally begin to develop a conceptual understanding of what happens when we add quantities if we help them recognize mathematics in familiar contexts. Money, for example, offers a familiar context to students, and reinforces the idea of place value. For example, 12 cents can be represented as 12 pennies or as 2 pennies and 1 dime. Classroom games that involve using number cubes (dice) to count or combine amounts also provide opportunities for students to practice adding and solidify their understanding of place value. Such practical contexts help students develop an understanding of early addition concepts, and are much more engaging than addition worksheets. When students are engaged in classroom activities that mimic activities they engage in outside of school, they usually concentrate on problem solving and do not necessarily think about the practice they are getting adding numbers. It is the less interesting paper-and-pencil desk work that gives rise to student disengagement, and may ultimately lead to classroom management issues.

Given the emergence of technology such as computers and calculators, which can easily be used to perform calculations, some confusion exists within the education community as to how many addition facts students should know by rote. For example, teachers may want students to immediately recall that $3 + 4$ is 7, but not necessarily that $14 + 27$ is 41. There is some general agreement that it is important for students to know addition facts up to 20 as a foundation; however, these facts should emerge naturally through exploration, and should be reinforced through activities and games. Students should learn these facts through contexts and practice-with-a-purpose, rather than by methods that promote rote memorization achieved through drill and skill activities. Punishing students who do not know the rote facts presented on flash cards only breeds stress, avoidance, and a dislike of mathematics. It is much better to provide multiple opportunities for addition practice that are fun, rewarding, and connected to the world outside of school. Online games may also provide options for students to practice their adding skills in an interactive way. For example, Kamii (2004) has developed many classroom games specifically designed to help students learn arithmetic facts and has demonstrated their usefulness in her research.

When it comes to recalling facts, speed is less important than understanding. Too many teachers still play math games in their classrooms that reinforce speed, but do nothing to help the students who are struggling—the kids who can recall

> **CONNECTION**
>
> For strategies for developing basic facts, see section 5.8.

their facts the fastest keep winning, and the kids who do so more slowly keep losing. Such games lead some students to compare themselves very negatively to others, making them dislike mathematics even more and feel even worse about their own abilities. Rather, students need to get faster at determining *strategies* to calculate unknown facts from known ones. Fluency will then naturally follow.

4.2 Addition Models, Strategies, and Procedures

Given the opportunity, students will quite naturally draw circles, sticks, or other marks to represent numeric quantities. They tend to begin by drawing more detailed pictures when using them to count objects; later, they shift from carefully drawn stick people or cookies to tally marks or drawings of simple circles to represent counters. This is the process of *mathematical abstraction* at work; students begin to realize that it is, for example, the *number of people* that matters in a given problem, rather than the specific details about the people. Addition is the most straightforward of the four fundamental operations to model since all of the numbers in a problem can be initially represented before it is necessary to think too deeply about carrying out the *operation*. For example, to model $3 + 4$, we can begin by representing both the 3 and the 4. As we will discuss in chapter 5, if the task were 3×4, representing both the 3 and the 4 would be less helpful.

There are many possible strategies that can be used to solve addition problems. Research has shown that if students are given appropriate problems, they are very capable of inventing addition strategies without direct guidance (see, for example, Kamii, 2004). As always, they should be provided with contexts to which they can relate, and problems that are meaningful to them. They should be given a choice of hands-on materials, and encouraged to use these materials to work out their ideas. When they are ready, children will become more comfortable representing numbers (or quantities) symbolically. Thinking and problem solving, rather than merely calculating correct answers, are the most important aspects of constructing meaningful understanding. Students should be given opportunities to share and explain their thinking and defend their strategies in small groups and in whole-class discussions. The teacher's role in such discussions is to probe and ask for clarification as needed; the teacher should not present him or herself as the sole authority on the ideas, or promote only one method. For this reason, teachers must be familiar with multiple methods. With the teacher's guidance, students should be the ones who ultimately determine whether or not the ideas are reasonable. If most or all of the students in the class are unsure that a solution is reasonable, then it is time for the teacher to rethink the problem, the context in which it was presented, and the teaching strategy used. Presenting an alternate context is more likely to help students construct meaningful understanding than simply providing the right answer. Once again, the student's ability to develop strategies, rather than the speed with which he or she can determine the correct answer, needs to be the focus of teaching and learning.

Exploration/Task

Students often come up with unexpected and insightful solution methods. Consider the following addition problem, followed by sample student solutions. Examine each student method, and try to determine the student's reasoning in each case.

Fred had 137 hockey cards in his collection. He got 4 more cards for his birthday, along with a package of 6 that came with a fast-food meal. How many cards does he have now?

Solution A:
$$137 + 4 = 137 + 3 + 1 = 140 + 1 = 141$$
$$141 + 6 = 147$$

Solution B:
$$137 + (4 + 6) = 137 + 10 = 147$$

Solution C:
$$140 + (4 + 6) = 150$$
$$150 - 3 = 147$$

Consider the following questions. If possible, discuss with a colleague.

- How might you have chosen to do the problem?
- How does your method compare with the solutions provided?
- Which method might be faster to perform?
- Which method(s) might allow you to more easily detect students' misconceptions?
- Which method teaches more about how numbers are composed and the meaning of addition and its properties?
- What if students don't come up with the traditional "carry the one" (regrouping) method on their own? Should you teach it to them?

Follow-Up and Discussion

You might have found it challenging to understand some of the student solutions provided in the task section; however, each is based on reasoning and understanding. In Solution A, the key idea is that if the 4 is broken up into 3 + 1, the 3 is easily added to 137. The 1 is then added, followed by the 6. In Solution B, the insight is that 4 + 6 give 10, which is also an easy number to add to 137. This idea is also used in Solution C, except the 137 is adjusted to the easier number 140 by temporarily adding 3. At the end, the 3 is subtracted again to bring the solution back to the correct value. All of these methods use insight to make the computations easier, and show a flexible understanding of numbers and operations.

The common classroom manipulatives called base ten blocks or centicubes (introduced in chapter 3) are usually readily available in schools, and are one possible model of decimal (meaning base ten) whole numbers. Use of such concrete models may help students explore and understand basic operations. For example, a student can model combining 10 ones when adding by trading 10 centicubes for a ten or long; this 10 is then recorded by placing a 1 in the tens column.

Practice and Further Exploration

Using the samples of student addition methods as a guide, find at least three different ways to think about and write the solution to the problem 189 + 23. Ideally, invent your own methods.

4.3 Subtraction: Concept and Contexts

While addition involves combining amounts, subtraction involves *removing* an amount (or amounts) from another; it is about what is remaining. Subtraction can also be used to compare two quantities by determining the amount (or the difference) between the two values. For example, we can compare the number of degrees between two temperatures as it gets warmer or colder, or the difference between the amount of money given to a sales clerk and the cost of an item, or the difference between the number of people in a room and the number of chairs available. When modelling addition problems, we modelled each value individually. When we are modelling whole number subtraction problems, we sometimes model only the first number. The number being subtracted is often *part of the first value*. For example, to model the problem 5 − 3, we might draw 5 circles and cross out 3 of them. Thinking about it this way, subtraction involves removing a portion of an initial value. Alternately, subtraction can be thought of as comparing two amounts using a one-to-one correspondence between two quantities and then seeing what is left. For example, if there were 5 cookies for 3 people, we could draw connections between individual people and cookies to determine that we are 2 cookies short. Unlike addition, the order of the numbers matters in subtraction: while it is true that 4 + 6 = 6 + 4, it is not true that 4 − 6 and 6 − 4 are equal.

While both interpretations of subtraction—removing and comparing—described above can be modelled, they apply to very different contexts. Students can develop misconceptions if proper contexts are not provided and distinctions between addition and subtraction are not explored. An example of such a misconception might be a student who feels that an appropriate strategy is to always subtract the smaller number from the larger. Though this is usually true in simple problems, students will later learn that this isn't always the case.

Everything that has been said about the importance of providing addition contexts (see section 4.1) is equally true for subtraction. Students need multiple opportunities to explore principles of subtraction using familiar contexts. They also need to be provided with materials that allow for hands-on exploration of subtraction scenarios. Problems that involve money and making change provide obvious, and generally highly familiar, contexts. Other contexts might include game scores, distances, differences in ages of siblings, and so on.

> **INSIGHT**
>
> The order of the numbers matters in subtraction.

Realistic contexts also help students understand that the value, or quantity, being removed is *part* of the initial quantity, rather than an additional quantity, or, alternately, that we are comparing two distinct values. In the latter case, it may be appropriate to model both values and then look for how much larger one is than the other—as long as students realize this is what they are doing and why.

Example

Consider the following problem:

> *Sue's mom made a batch of 12 brownies last night and said it was okay for Sue and her little brother, Johnny, to eat some after school. When Sue got home, she counted the brownies and there were only 5 left. Meanwhile, Johnny and his best friend were running around the family room. How many brownies might the boys have eaten?*

One approach to modelling this problem might be to represent *both* the 12 original brownies and the 5 brownies that remained when Sue got home; however, rethinking the context makes it clear that the 5 brownies *are part of the 12*. In fact, we need only represent the 12, and then identify 5 *of these*. Here is one possible way to model this problem:

There were 12 brownies ...

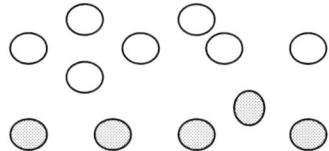

and now there are only 5 left.

Modelling subtraction differs from addition in that we often do not initially represent each of the amounts specified in the problem. For example, we would probably not solve the problem described above by initially drawing 12 brownies *and* drawing 5 brownies; rather, it is likely that we would draw 12 brownies and then *select* 5 of the 12. This concept is much easier to understand in a context than it is by using a rule. The ongoing use of contexts when solving subtraction problems is critical.

Subtraction can also be used to find out how far apart two quantities are; for example, how many degrees the temperature rises from a low of 5 degrees to a high of 12 degrees. This can be illustrated by looking at a thermometer. (Note that when turned horizontally, a thermometer clearly shows a number line—a horizontal model with numbers that increase from left to right.)

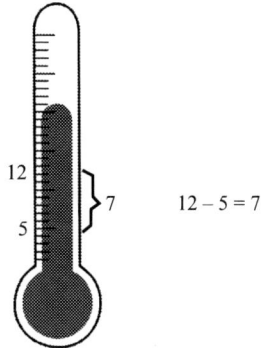

$12 - 5 = 7$

Sometimes, depending on the context, it helps to make a one-to-one correspondence between the two quantities and then see what is left. For example, what if we had only 5 chairs but 12 people came to our party? How many people do not have a chair? In this case, the model would look different from a model we would use to show that an amount has been removed from a quantity.

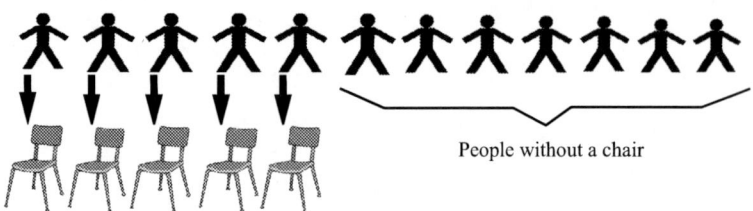

The one-to-one correspondence of people to chairs in this figure illustrates a comparison approach. Sometimes, emphasizing the connection between addition and subtraction is particularly helpful in such examples. For example, the brownies example provided above could also be expressed as ? + 5 = 12, in which case it is called a *missing addends problem*. In this case, students might think, "How much do I need to add to 5 to get 12?" This method could also be used to compare amounts; for example, students may ask, "How many chairs do I need to add to the 5 chairs I now have to allow all 12 people to have a seat?"

When students reflect on what they might do in a realistic situation involving subtraction, they have the opportunity to develop numeric strategies of their own *with understanding*. It is important to provide multiple contexts for students, as well as to regularly remind them of the connection between addition and subtraction. Once again, the concept of place value is critical. Students whose teachers tell them to use a formal rote method, especially those students who do not deeply understand place value, are typically at risk for producing solutions such as the one outlined in the following section.

Exploration/Task

Here is a student's solution to the sample problem 305 − 86.

$$\begin{array}{r} \overset{1}{\cancel{2}} \\ \cancel{3}\ ^{10}\ ^{15} \\ -\ \ 8\ \ 6 \\ \hline 1\ \ 2\ \ 9 \end{array}$$

Examine the student's calculation. What misconceptions might be at work? What problem could you pose to the student to determine if you have indeed identified their error? What models, manipulatives, and other examples might the student be provided with to help him or her move forward?

Follow-Up and Discussion

The student who provided the solution in the above example appears to be confused about the relative values of the digits in each column. He has attempted to regroup so that the 5 in the ones column is written as 15; however, he has regrouped a group of 100 (from the 300) directly from the hundreds column to the ones column. The new value in the ones column, 15, is really 105.

This example illustrates well how confusing the standard method can be. Invented methods can be much more meaningful, as well as less likely to lead students to make errors. For example, relying on the connection to addition, a student might solve the problem correctly by starting with 86, and adding numbers until 305 is reached, by writing

$$86 + 14 = 100$$

$$100 + 205 = 305$$

Using this method, the student arrives at the correct answer of 14 plus 205, or 219.

Allowing students to construct their own solution methods for problems involving familiar contexts by relying on modelling and reasoning usually prevents them from developing procedural misconceptions (such as the one that was evident in the preceding student example). As students further develop these conceptually based methods, numeric methods will naturally emerge, but formal procedures that are imposed before conceptual understanding is in place are often poorly understood and easily forgotten.

4.4 Subtraction Models, Strategies, and Procedures

As with addition, students may begin to model subtraction problems by using detailed drawings, but gradually their drawings will become more abstract and symbolic. As students work with larger values, they may eventually choose to use number symbols and either invented or more traditional numeric strategies. The work they have done previously to develop strategies using concrete materials and drawings will stand them in good stead as they begin to use symbols. Experience with manipulatives allows students to develop a deeper understanding of values and operations, which they can apply to number symbols in a more generalized way. Students may invent many effective strategies, such as rounding up to more manageable numbers (and adjusting later), working with left-most columns first (instead of working from right to left, as is done in some procedures), adding up (instead of directly subtracting), and so on.

Example

Consider the following subtraction problem, which is followed by some sample student solutions:

Mrs. Dees has bought a package of 75 chocolate bars to give to the students in her two classes. There are 28 students in the earlier class and each student took 1 bar. How many chocolate bars are left for the second class?

Solution A:

$$75 - 30 = 45$$

30 is subtracted as it is a round number that may be more manageable than 28. Since not quite that many chocolate bars were eaten, 2 must be added back on:

$$45 + 2 = 47$$

Solution B:
$$75 - 28 = (70 - 20) + (5 - 8) = 50 - 3 = 47$$

Here, the numbers are decomposed and subtracted by place value. Surprisingly, some children easily recognize that (5 − 8) is equivalent to subtracting 3.

Solution C:
$$75 = 60 + 15$$
$$60 + 15 - 20 - 8 = (60 - 20) + (15 - 8) = 40 + 7 = 47$$

Again, the numbers are decomposed, but this time using 60 + 15 to replace the 75 allows the calculation of 15 − 8 to have a positive result.

Solution D:
$$28 + 2 = 30$$
$$30 + 45 = 75$$
$$45 + 2 = 47$$

(See also the missing addends examples in sections 2.1 and 4.3)
In this case, 2 is added to 28 to yield the "nice number" of 30. The student then adds up from 30 to 75, finding that 45 is needed to do so. The 2 then needs to be added back on, as the initial value was 28.

Solution E:

$$\begin{array}{r} 7\ ^{1}5 \\ -\ ^{3}\cancel{2}\ 8 \\ \hline 4\ 7 \end{array}$$

Like the traditional method, this method uses an extra 10 with the 5 from regrouping; however, instead of borrowing from the 70, another 10 is subtracted along with the 20. In other words, 30, rather than 20, is subtracted to compensate for the extra 10 that was added to the 5.

Solution F:

$$\begin{array}{r} ^{6}\cancel{7}\ ^{1}5 \\ -\ 2\ 8 \\ \hline 4\ 7 \end{array}$$

This solution shows the traditional method of regrouping a 10 from 70 to give 15 in the ones place.

Exploration/Task

Examine each sample solution provided above and explore the conceptual basis for each method. Determine if the method is reasonable, and if it would work again in similar circumstances.

Follow-Up and Discussion

Those of us schooled in a traditional mode of algorithmic solutions might feel at some level that the regrouping method (Solution F) is the right one; however, for students who have not extensively practiced such a method, any or all of the other solution methods might seem reasonable (as may others that are not provided here). Allowing students to share and explain the reasoning behind their methods supports both a flexible understanding of numbers, and the development of an arsenal of efficient problem-solving methods. For example, when a student is comfortable thinking in this mode, solving a question such as 300 − 198 as 2 + 100 is not only faster, but is also less likely to lead to errors.

Practice and Further Exploration

Compute 301 − 184 the way you normally would. Now, apply each of the methods shown in sample solutions A through F to this problem. As you work, think about which method is the least prone to errors and promotes the greatest understanding of the composition of numbers and operations. Think about how a student might explain the method. Is the explanation conceptual or procedural? Does it imply understanding of the operation? Which (previously unfamiliar) method is your favourite? Resolve to use that method every time you need to subtract for the next month; at the end of the month reconsider which method works best for you.

4.5 The Connection between Addition and Subtraction

There is an intrinsic relationship between addition and subtraction.

Sometimes it is handy to use addition to solve a problem that suggests subtraction. This is not only acceptable, but should be promoted as it helps students explicitly develop the connection between the two operations. For example, solve the following problem in your head:

You have $2.05. The snack you want to buy costs $1.98. How much money will you have left after you pay for your snack?

Reflect on your thought process: Did you add or subtract to solve the problem? The ongoing use of context, and allowing and encouraging students to invent, share, and use their own methods has yet another advantage; learning this way also makes obvious the relationship between addition and subtraction. Mathematicians call addition and subtraction *inverse operations.*

For example, if you deposit $20 into your bank account, and then withdraw it again, your balance will return to where it was before you made the deposit. Symbolically, we could write B + 20 − 20 to represent the transaction, where B is your account balance. Subtracting the $20 reverses the initial deposit, and returns the balance to the amount that it was when we started.

It is much easier for students to understand the relationship between addition and subtraction using a context than it is if they are formally taught that these operations are inverse, which is not, in and of itself, all that exciting or interesting to most children. Meaningful contexts, as opposed to rules and definitions, help students build conceptual understanding.

INSIGHT

Addition and subtraction are related. They might be understood as combining or removing quantities, or distances on a number line. A subtraction operation can "undo" an addition operation, and vice versa.

CONNECTION

Addition and subtraction can often be used in place of one another (see sections 2.1 and 4.4).

MATHEMATICAL TERM

Inverse operations are operations that undo or reverse each other.

4.6 Connecting to Multiplication Concepts and Contexts

Addition and multiplication are also intrinsically related to each other—one way to think about multiplication is that it is one amount added repeatedly. The idea of making groups of an amount is *multiplicative thinking*. These groups could be many, few, or even partial. We might think of 4 packages that each contain 3 golf balls as $3 + 3 + 3 + 3$, or as 4×3, in which case the "$4 \times$" can be thought of as "4 groups of." Consider the following problem, which is followed by the initial part of three different student models of the situation:

> *Maria wanted to become more fit, and decided to start slowly. For the first week, she decided to do 5 sit-ups at a time, 3 times a day. How many sit-ups did she do each day?*

Student model A:

Student model B:

XXXXX XXXXX XXXXX

$5 + 5 + 5$

Student model C:

√ √ √ √ √
√ √ √ √ √
√ √ √ √

Exploration/Task

Examine each student's initial model. Can you determine if each student understands multiplicative reasoning? Which model has the best potential to illustrate the meaning of the multiplication operation? What feedback or follow-up tasks would you provide next for each student?

Follow-Up and Discussion

Student A has begun the problem by drawing both numbers, likely before thinking about what the problem is really asking. This might reflect *additive* reasoning, as they are perhaps thinking that the numbers 3 and 5 will be directly combined in some way, as might be done in addition. This student may or may not be ready to learn about multiplication. Student B on the other hand, seems to understand the context of the problem, and the role of the number 3. The 3 could be used to tell us that we have *3 groups of 5*—this is *multiplicative reasoning*, although the student then adds the groups of 5 using repeated addition (which is equivalent to 3 × 5). Student B already understands the basic concept of multiplication and is ready to learn more. Student C clearly understands the role of the 3 and the 5 in the problem and has provided a model called an array, a two-dimensional grid of check marks, to illustrate the answer, which shows multiplicative reasoning. Array models will be further explored in chapter 5.

Chapter Problems

1. A baker makes 53 chocolate chip cookies and 26 oatmeal cookies. Describe two mental math strategies that could be used to determine how many cookies the baker made in total.
2. A baker makes 58 cookies. Three of his customers buy 12 cookies each. Describe two mental math strategies that could be used to determine how many cookies remain after these purchases.
3. A baker makes 58 cookies. He needs 12 bags, each of which contains an equal number of cookies. How many cookies are in each bag? Draw a picture to represent this scenario and describe how it differs from a picture or model you would have drawn to represent Problem 2. How does the answer, the meaning of the answer, and the remaining number of cookies compare in each case?

Further Reading

Lawson, A. (2007). Learning mathematics vs. following "rules": The value of student-generated methods. *What Works? Research into Practice* (Research Monograph #2), 1–4.

> Lawson describes errors that students often make when adding and subtracting, and discusses allowing students to invent their own strategies for algorithms and how teachers can support this.

Thomas, J. N., and Tabor, P. D. (2012). Developing quantitative mental imagery. *Teaching Children Mathematics, 19*(3), 174–183

> Thomas and Tabor discuss early arithmetic concepts for Grade 1 students and their understanding of these concepts. The article includes ways to support their development.

Voza, L. (2011). Winning the "Hundred Years' War." *Teaching Children Mathematics, 18*(1), 32–37.

> Voza describes strategies that can be used to help students subtract with regrouping by moving from the concrete to the abstract.

Chapter 5
Whole Number Multiplication and Division

5.1 The Connection between Addition and Multiplication

There are a number of possible interpretations of multiplication, but repeated addition is a useful starting point.

However, there is a critical conceptual leap between additive reasoning and multiplicative reasoning. Multiplicative reasoning is based on repeated groups or amounts, while in addition operations both (or all) numbers play the same role. In multiplication, the *repeated addition* interpretation involves understanding one value as a quantity or amount and the other as the *multiplier*. When we think of doing something 3 times a day, the 3, the multiplier, *operates* on the other number (and this other number represents the *quantity* that is being repeated). Doing 5 sit-ups 3 times a day can be represented as an amount or quantity of 5 and a multiplier that tells us how many groups of that 5 there are (in this case, 3). This operation can be written numerically as repeated addition: 5 + 5 + 5. As a multiplication problem, it can be written as either 3 × 5 or 5 × 3, but, in each case, the 5 represents a concrete amount and the 3 relates to the *operation on* the amount. Each of the expressions above calculates the same amount (we could even have written 2 × 5 + 5), but in each case, the role of the 3 is different than when we are *adding* 3 to another value. There are also ways to conceptualize multiplication other than repeated addition, as will be discussed below.

> **CONNECTION**
>
> See section 4.6 to review repeated addition as preparation for multiplication. This involves the idea of making groups of a number.

5.2 Multiplication Models, Strategies, and Procedures

This section explores ways to model multiplication.

Exploration/Task

Consider the following problems. Draw a simple model to represent the situation presented in each problem.

Problem A:

Two students each bring a package of 6 juice boxes to share at school. How many juice boxes were there to share?

Problem B:

Six children each took off both of their socks and put them in a pile. How many socks were in the pile?

Follow-Up and Discussion

Although the numbers used in these problems are the same, the models used to represent them may differ. For Problem A, you might have drawn 2 groups of 6, while for Problem B you might have drawn 6 groups of 2. While it might be obvious to us, as teachers, that the answer to each problem will be the same, reaching this conclusion requires some experience with multiplication. From a modelling perspective, 2 groups of 6 and 6 groups of 2 are *not* the same. Students will thus need to explore other interpretations of multiplication before it becomes clear that we can multiply in any order. Consider the example 2×3. If we think of 2 groups of 3, we might draw:

◯ ◯ ◯   ◯ ◯ ◯

Depending on the context, students might also interpret 2×3 as 2 items, repeated 3 times:

◯ ◯   ◯ ◯   ◯ ◯

Another model of multiplying is called the *area model*. In an area model, two numbers being multiplied are represented as side-lengths of a rectangle: one value is the number of rows, while the other is the number of columns of the rectangle. The *area* of that rectangle—as represented by the number of squares within the rectangle with side-lengths of 1 (called unit squares)—represents the answer. The area model below represents 2×3. Could this model also be used to represent 3×2?

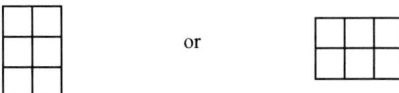

> **MATHEMATICAL TERM**
>
> *Equivalent* means having the same value. For example, $\frac{1}{2}$ and $\frac{2}{4}$ are equivalent fractions; 2×3 and 6×1 are equivalent expressions.

Using the area model to count the number of unit squares (and find the answer to the multiplication problem) immediately makes it obvious that 3×2 and 2×3 are *equivalent*.

The area model is very useful because it makes it visually clear that order does not matter in multiplication. This property of multiplication is called the *commutative property*.

> **MATHEMATICAL TERM**
>
> An operation is *commutative* if it can be performed in any order. Addition and multiplication are both commutative.

> **CONNECTION**
>
> To review the commutative property for addition, namely the idea that numbers can be added in any order, see sections 4.1 and 4.2.

Another model for multiplication, which is conceptually similar to the area model, is the array model. The above example could also be illustrated using an array model.

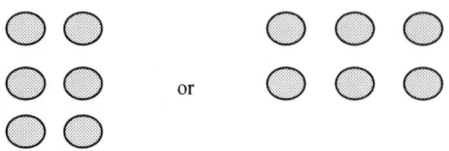

The area model and the array model clearly show that numbers can be multiplied in *any order*. It is not necessary to multiply numbers from left to right, or in any specified order or grouping. If students are introduced to the area model, they will be able to recognize this important property themselves.

The area model also allows students to make a direct connection between multiplication and two-dimensional area. Using this type of model, it is easy to see that a room that is 2 metres by 3 metres has 6 *square* metres of space; it is also clear that we need a new unit—square metres rather than metres—to account for this space. A diagram shows how reasonable it is that a 1-metre-by-1-metre square is called a *square metre*.

> **CONNECTION**
>
> Multiplication using an area model sets the stage for learning to calculate and measure area (see section 14.2 for more detail on calculating area).

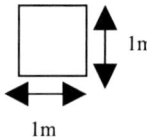

Students might invent other multiplication models depending on the situation. Consider the following problem:

Two sisters each own a grey, a black, and a navy skirt. How many skirts do they own in total?

A model such as this could be used to depict the problem:

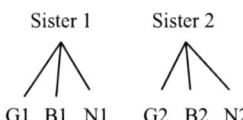

This example illustrates multiplication as a number of combinations—at a later stage, in algebra, these may be referred to as *a cross product of sets*.

An important aspect of modelling is to help students make the connection to numeric methods that could be used to solve the problem, and the area model is helpful for this.

> **CONNECTION**
>
> Counting possible outcomes is also used in probability. For applications of this type of model, sometimes called a tree diagram, see chapter 15.

Example

Consider the calculation 13 × 25. Just as we modelled 2 × 3 by creating a 2-by-3 rectangular area, we can model 13 × 25 with a 13-by-25 rectangular area:

The model also allows us to see the numbers 13 and 25 in another way. Each can be grouped by place value, as 10 + 3 and 10 + 10 + 5 respectively:

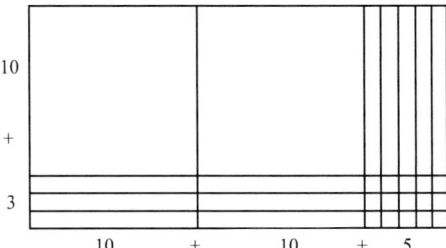

Exploration/Task

Use base ten blocks to build the model for 13 × 25. If you don't have blocks, redraw the area model on paper and write in the value of each base ten block piece as you draw it. For example, the larger 10-by-10 square, called a *flat*, represents 100. Once you have built (or drawn) the model, begin adding up the areas, which represent the *product* of 13 × 25. Think about how you might be able to find the product value using (only) your model, before reading on.

MATHEMATICAL TERM

A *product* is the answer to a multiplication problem.

Follow-Up and Discussion

Let's examine the area model for the 13 × 25 example in more detail, working to connect it to the construction of a numeric method for multiplication. Compare your own drawing to the following model, which groups the area of the rectangle into sub-regions, or sub-rectangles. Compare the dimensions of each sub-rectangle to its area.

	20	5
10	10 × 20 = **200**	10 × 5 = **50**
3	3 × 20 = **60**	3 × 5 = **15**

For example, in the largest rectangle in the model, the 10 and the 20 are the dimensions of the rectangle, and 200 is the space generated, or the product. The 10 and the 20 represent the tens digits of each number being multiplied in 13 × 25.

Exploration/Task

Using your model, examine the other three sub-rectangles, linking the side-lengths of each sub-rectangle to the numbers (10 and 3, and 20 and 5) in the original problem and model. Think about how the model and these numbers might connect to generate a numeric method.

Follow-Up and Discussion

We can add up the areas of these sub-rectangles in any order to find the answer or product.
For example, we can add the areas *horizontally*, and then vertically:

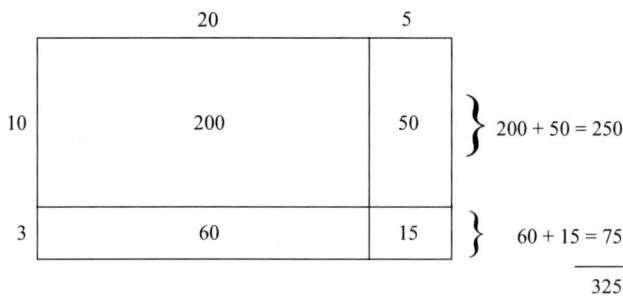

or *vertically* first, and then horizontally:

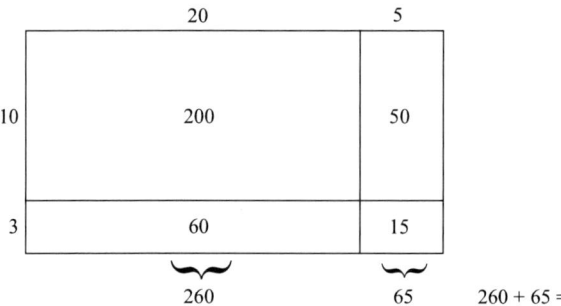

Using the area models, we see that the answer to the problem 13 × 25 can be obtained by adding up the four numbers (indicating area) in the models *in any order;* thus, it seems reasonable that it would be possible to use a number of different numeric methods to calculate the area. For example, we can start with the largest rectangle and add the regions from largest to smallest:

$$
\begin{array}{r}
13 \\
\times\ 25 \\
\hline
200 \\
60 \\
50 \\
15 \\
\hline
325
\end{array}
$$

If we add the areas in a different order, the method might begin to look more familiar to some teachers (although not necessarily to students):

$$
\begin{array}{r}
13 \\
\times\ 25 \\
\hline
15 \\
50 \\
60 \\
200 \\
\hline
325
\end{array}
\qquad
\begin{array}{r}
1^13 \\
\times\ 25 \\
\hline
65 \\
260 \\
\hline
325
\end{array}
$$

The method shown on the right is sometimes called the *traditional method*. It simply collapses two of the sums obtained from the area model; however, carrying or regrouping the numbers to the next column can often lead to careless procedural errors that students are more likely to avoid if they write out each value separately, as is shown in the example on the left. Again, it should be emphasized that the values from the area model can be added in any order.

Students are likely to invent a variety of numeric procedures when given the opportunity to do so. In the Exploration/Task to follow, you are asked to investigate and determine the validity of a selection of student-generated strategies or methods.

Exploration/Task

Examine each solution to the problem provided and explore its conceptual basis and validity. Try to use each student method to solve a similar problem of your choice.

There were 17 players on each hockey team in the league. There are 6 teams in the league. How many players were there in total?

Solution A:

To calculate 17×6, we know $17 = 15 + 2$

$$15 \times 6 = 15 \times 2 \times 3 = 30 \times 3 = 90,$$

$$\text{and } 2 \times 6 = 12$$

$$90 + 12 = 102 \text{ players in all}$$

Solution B:

$$17 \times 6 = 34 \times 3 = 90 + 12 = 102$$

Solution C:

Adjust 17 to 20:

$$20 \times 6 = 120$$

Subtract $3 \times 6 = 18$:

$$120 - 18 = 102$$

Follow-Up and Discussion

In Solution A, the student uses the idea that 17 groups of 6 can be thought of as 15 groups of 6 and 2 groups of 6. Solution B demonstrates a flexible thinking strategy, doubling 17 and halving 6, to create an equivalent but easier problem. For Solution C, the student uses 20, a round number that is easier to deal with, instead of 17, and then adjusts the final answer.

We saw earlier that the two-dimensional area model can be useful when multiplying two numbers. Similarly, there is a direct connection to be made between multiplying three numbers and three-dimensional space (volume).

> **CONNECTION**
>
> Multiplication models can also be used to understand and calculate volume (see section 14.6 for an in-depth discussion of this concept).

Exploration/Task

If you have access to centicubes or interlocking cubes (which actually snap together), build two separate 3-cube-by-4-cube layers, each in a different colour.

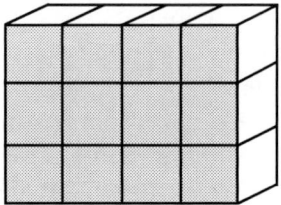 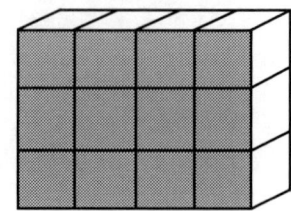

In total, there are 3 × 4 cubes in each of the two rectangular layers. Stack one rectangular layer on top of the other. How can you find the total number of cubes used in the new shape?

Follow-Up and Discussion

The model allows you to see that there are still 3 × 4 cubes in each layer, times 2 layers. One way to figure out the total number of cubes is to determine the number of cubes in the bottom layer, then multiply by the number of layers. Numerically, this is expressed as: (3 × 4) × 2 = 24 cubes. (If you use centicubes, each cube represents 1 *cubic* centimetre in volume, which is a cube in which all sides measure 1 centimetre in length.)

Look again at your block of 24 cubes. What if we sliced it vertically like a block of cheese? We might have four 3-by-2 slices (still 24 cubes in all). Or if we sliced it lengthwise, we would have three 2-by-4 slices. Try this with your model. The block or cube model illustrates another property of multiplication called the *associative property*.

> **INSIGHT**
>
> We can pair up and multiply numbers in any order in a multiplication statement.

Using the cube model, we can see that both of the following ways to calculate the total number of cubes are equivalent:

$$(3 \times 4) \times 2$$

$$3 \times (4 \times 2)$$

and if we also use the commutative property, both of these ways are also equivalent to (3 × 2) × 4, as was illustrated by the block of cheese analogy.

Familiarity and fluency with these properties helps to develop a deeper understanding of the multiplication operation and, therefore, helps with computation. For example, the calculation 17 × 5 × 2 appears difficult, unless we see the potential for the associative property: 17 × 5 × 2 = 17 × (5 × 2) = 17 × 10 = 170. This calculation is so much simpler than working from left to right.

Exploration/Task

Make use of the properties of multiplication to calculate 50 × 23 × 2 in the easiest way.

Follow-Up and Discussion

It is important for teachers to help students be on the lookout for situations in which they can apply flexible thinking, such as finding 50 × 2 first (and then multiplying it by 23) in the previous Exploration/Task.

5.3 Division: Concept and Contexts

> **INSIGHT**
> Division can mean splitting something into groups.

Equal sharing is often one of the earliest contexts in which children encounter division. A young child with 2 cookies knows that when she is told to "share with Sam," she will keep 1 cookie and give 1 to Sam. This is analogous to thinking of 2 cookies divided into 2 equal portions, which results in one cookie each, or $2 \div 2 = 1$. This equal sharing understanding of division is sometimes called the *partitive model of division* as it involves partitioning (splitting up) an amount into equal parts.

> **MATHEMATICAL TERM**
> The *partitive model of division* is enacted by splitting a quantity into equal groups.

Exploration/Task

Consider the following problem context:

> *A bag of 20 slices of bread was used to make sandwiches. If 2 slices of bread were used per sandwich, how many sandwiches could be made?*

Model and solve this problem. Think about what operation you are using.

Follow-Up and Discussion

In your solution, you might have found yourself thinking in terms of division (for example, $20 \div 2$), or you might have been thinking in terms of subtraction ($20 - 2 - 2 - 2...$ or "How many twos before I use up all the bread?"), or even in terms of multiplication (for example, "2 times what number is 20?").

> **CONNECTION**
> To review division and multiplication, and the relationship between them, see section 5.4.

The repeated subtraction idea is, indeed, another way to think about division. It is a very important model of division especially for non-whole numbers.

This *repeated subtraction* idea is called the *measurement model of division*, because it entails counting out or measuring out portions (in this case, 2 slices of bread for each sandwich).

> **INSIGHT**
> Division can be thought of as repeated subtraction.

It is very important that students learn about *both* the partitive and measurement models of division when working with whole numbers, as well as the connection between multiplication and division, so that they develop a deep and flexible understanding of the process. The measurement model in particular is important when students begin to work with fractions. Although it is not necessary for students to learn the names of these models, teachers need to ensure that students have the opportunity to examine multiple types of division problems with whole numbers, particularly before introducing fractions.

> **INSIGHT**
> Division can be thought of as measuring or counting out portions.

Exploration/Task

Model and solve each of the problems below. Determine which interpretation of division (partitive or measurement) is being used in each case.

Problem A:

> *There were 12 cookies to be sold at the bake sale. If these cookies were wrapped in packages of 3, how many packages of cookies were for sale?*

Problem B:

> *There were 12 cookies to be shared among 3 children. How many cookies did each child get?*

> **MATHEMATICAL TERM**
> The *measurement model of division* is enacted by counting or measuring out portions through repeated subtraction.

Follow-Up and Discussion

You should have reached the same final answer as you created the model and solved each problem above; however, your model and interpretation were likely different in each case. You may have modelled Problem A by drawing the packages of 3, one at a time—until all 12 cookies had been place in a group. You would find that you could make 4 groups of 3. For Problem B, you may have begun by handing out the cookies (almost like you were dealing cards) to each of the 3 children. When the cookies were gone, you would have noticed that you had given 4 cookies to each child.

It is also possible to think through a division question by using what we know about multiplication, as we will explore in the next section.

Practice and Further Exploration

Draw a model to depict each context described below and use the model to solve the problem. Identify the interpretation of division (as described in this section) that you have used in each case. How do your interpretations of the numbers differ?

Problem A:

> A particular game has 24 tokens that are to be shared equally among the players. If there are 6 players, how many tokens does each player get?

Problem B:

> At a sports event, a coach has 24 drink coupons. She gives 4 each to each athlete in a group of players. If each player received the same number of coupons, how many players are in the group?

5.4 The Connection between Multiplication, Division, and Ratio

In chapter 4 we illustrated how addition and subtraction are connected, and thus how one of these operations can sometimes be used to solve a problem involving the other. Multiplication and division also have a similar intrinsic relationship, which is why it can be useful to compare approaches to problems that could be solved using either operation. The multiplication model of repeated addition is yet another example of a connection between operations, which in turn relates to the division model of repeated subtraction. While the operations were historically treated as separate topics in the curriculum, it is the potential for developing connections between them that contributes to deep understanding. Let's examine a question from the previous section to look for the connections among methods.

To solve the problem of the 24 tokens to be shared equally among 6 players, we need to find out how many tokens each player will receive. We might solve this using any one of the following methods (or others):

Division:

$24 \div 6 = 4$

Multiplication:

6 × ? = 24; knowing 6 × 4 is 24 allows us to find the answer of 4

A physical model of 6 piles:
24 tokens are given out one at a time until all of the tokens have been given out; then the number in each pile is counted.

A guess-and-check strategy:
Give 2 tokens to each player. You have distributed 2 × 6 or 12 tokens. Note that you have distributed half of the tokens. You now know each player needs 2 *more* tokens, and can distribute 2 × 6 or 12 tokens once more. Each player ends up with 2 + 2 or 4 tokens

Repeated addition:
Adding 4 + 4 + 4 + 4 + 4 + 4 until you reach 24.

Repeated subtraction:
Subtracting 24 − 4 − 4 − 4 − 4 − 4 − 4 until you reach 0. You can see that 6 amounts of 4 were subtracted from 24.

Yet another way to think about the token problem is to use a ratio. A *ratio* is a comparison between two amounts. For example, the *ratio* of tokens to players in the example is 24:6. In this case, finding a *unit ratio* means finding how many tokens there are for *1* player. We might write 24:6 = ?:1 or 24 ÷ 6 = ? ÷ 1. The model below can be used to think about the ratio relationship, and also emphasizes the connections among operations. The ratio model connects multiplication and division, as well as the idea of ratios (including unit ratios) and fractions.

> **MATHEMATICAL TERM**
>
> A *ratio* is a comparison of two quantities of the same unit. It can be expressed as a comparison (3:4) or a fraction $\left(\frac{3}{4}\right)$.

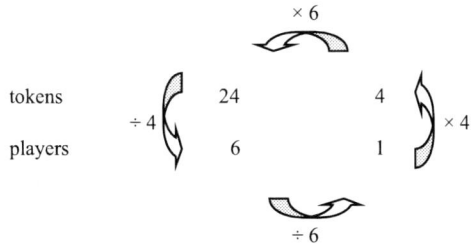

If we look at the vertical arrangement of the numbers in the model, on the left side we see the ratio idea of 24 tokens for 6 players. This implies the unit ratio of 4 tokens per player, which is visible on the right side. Looking at the model this way (vertically) shows the partitive model of division, that is, division as equal sharing.

On the other hand, when we look at the model horizontally, we see the equivalent problem of how many people will get tokens if there are 24 tokens and they are given out 4 at a time. Looking at the second horizontal row, we see that 6 players each get a portion. The arrows show the relationship between multiplication and division, thus the ratio model may be helpful in showing students how the operations link together.

Exploration/Task

Consider the following sample problem:

> *Mom bought a bag of 72 treats for the party. There were 12 children at the party. How many treats should each child get?*

Solve this problem two ways—one that uses division and one that uses multiplication. Then sketch a ratio model (as illustrated above) and compare it with your solutions.

Follow-Up and Discussion

When solving the above problem you might have written:

$$72 \div 12 = 6$$

and

$$12 \times ? = 72$$

We notice that several related calculations arise from the given context, including $12 \times 6 = 72$ and $72 \div 12 = 6$. The problem can be modelled by dealing out the candies into 12 piles until they are gone, resulting in 6 candies in each pile. It is also possible to write $72 \div ? = 12$. A student who knows that $12 \times 6 = 72$ might know that this problem could be solved by making piles of 6 treats, then confirming that there are 12 piles.

Inverses

Multiplying by any given number, then dividing by the same number has a result that is similar to adding and then subtracting the same number; hence, these operations are called *inverses*. Intuitively, inverse operations undo each other. For example:

$$6 + 4 - 4 = 10 - 4 = 6$$

$$5 \times 4 \div 4 = 20 \div 4 = 5$$

Here we calculated $5 \times 4 \div 4$ from left to right, but we also notice that $4 \div 4$ is 1. If we could calculate the "$\times 4 \div 4$" part of the operation first, it is obvious that without any further calculating "$\times 4 \div 4$" does not change the initial value of 5 since it is 1. In the following section, we will explore whether this is a reasonable step to take.

Exploration/Task

Calculate each set of expressions, first by working from left to right, and then by calculating the bracketed expressions first. For each set of calculations, determine if the answer is the same when it is found using the varying order shown with the brackets. In other words, explore whether multiplication and division can be grouped and performed in any order.

1. a) $21 \times 7 \div 7$
   b) $21 \times (7 \div 7)$
   c) $7 \times (21 \div 7)$
2. a) $3 \times 12 \div 12$
   b) $12 \div 12 \times 3$
   c) $3 \times (12 \div 12)$
   d) $12 \times 3 \div 12$

> **INSIGHT**
>
> Multiplying by a number, and then dividing by the same number takes us back to where we started. This is because multiplication and division are *inverse operations*.

Follow-Up and Discussion

These examples demonstrate that we can indeed pair both multiplication and division operations and perform them in any order. A key idea is that the operation sign must remain with the number it precedes. For example, in $12 \times 3 \div 12$, we are always multiplying by 3, regardless of the order in which the operations are performed (we cannot change to dividing by 3). Multiplying and then dividing again by the same number (with the exception of 0, which is addressed in chapter 10) does not change a given value.

> **MATHEMATICAL TERM**
>
> The idea that we can group multiplication and division operations and perform them in pairs in an order other than left to right is called the *associative property*.

5.5 Division Models, Strategies, and Procedures

In section 5.3, a number of different interpretations of division were illustrated, with the choice of interpretation depending on the context. The two most important interpretations, the partitive and measurement models of division, are once again illustrated in the following problems.

Exploration/Task

Draw one model to illustrate each of the situations described below. Compare the two models, and the contextual interpretation of the answer in each case.

Problem A:

A spool of 10 metres of twine is available for tying boxes. About 2 metres of twine are required to tie each box. How many boxes can be tied?

Problem B:

There are 10 tennis balls in the bin, to be shared equally between 2 teams. How many tennis balls does each team get?

Follow-Up and Discussion

Problem A can be solved by subtracting or measuring out 2-metre lengths until all of the twine is gone. Recall that this is called the *repeated subtraction* or *measurement model*. Problem B involves separating or partitioning the balls into 2 groups or parts; hence it uses the *partitive model*. In each problem, we are dividing by 2; however, in Problem A, 2 is the *size* of each group or part, and in Problem B, 2 is the *number* of groups or parts.

The Traditional Procedure

As with the other arithmetical operations, teachers are cautioned against imposing formal numeric procedures too soon—if at all. The traditional procedure sometimes called *long division* is particularly risky. This procedure has many steps and rules, and does not connect easily to a concrete model. Often, teachers do not connect language that reinforces place value to this particular algorithm—making it one more procedure for students to memorize.

Exploration/Task

Compare the two student approaches shown below. In each case, the numeric method is shown, as is the student's initial discussion with a teacher.

Student A:
Elaine describes how she started to divide 264 by 12 using the long division procedure.

$$\begin{array}{r} 2 \\ 12 \overline{\smash{)}264} \\ 24 \end{array}$$

Elaine: I started with 26 so I wrote a 2 on top.
Teacher: Why did you write a 2?
Elaine: Because 12 goes into 26 twice.
Teacher: And then what did you get when you subtracted 24 from 26? …

Student B:
Donna begins the same problem as Elaine, but uses the repeated subtraction method, which involves subtracting groups of the number by which we are dividing.

$$\begin{array}{r} 12 \overline{\smash{)}264} \quad\; \\ 120 \;\big|\; 10 \end{array}$$

Donna: I wasn't sure how many groups of 12 there were in 260, but I thought there had to be at least 10 so I started with that.
Teacher: How did you know there had to be at least 10 groups?
Donna: Because 12 × 10 is 120. So then I subtracted 120.
Teacher: Remind us why you subtracted the 120?
Donna: Because I had already divided out those 10 groups of 12, which takes care of 120.…

Follow-Up and Discussion

Elaine is using a procedural method. The use of the 26 and the 2 in Elaine's description suggests she doesn't deeply understand the reasoning in the method; in fact, these values actually represent 260 and 20. The teacher's line of questioning supports only the enactment of the procedural steps, rather than probing the reasons behind them. The teacher should have questioned Elaine's use of the number 2, as this value really represents 20.

Donna is already using a method that is more likely to be based on understanding. Indeed, the use of the numbers 260 and 10 suggests an awareness of why the method is working. The teacher supports that awareness by ensuring that everyone understands that the method involves repeatedly subtracting groups of 12 and counting how many groups of 12 there are in all. Using a procedure without really understanding it is risky. In addition to being prone to forgetting the method, students may not really understand what the answer means.

Exploration/Task

Consider the following problem, and the sample student solutions provided below:

There are 180 students and 3 teachers going on a field trip. Buses carry 44 passengers each. How many buses are needed for the 180 students and 3 teachers to attend?

Sample solution A:

$$44 \overline{)183} \\ \underline{-176} \\ 7$$ (quotient 4)

The answer is 4 remainder 7. The answer is 4 buses.

Sample solution B:
183 passengers. Try 3 buses:
$3 \times 44 = 120 + 12 = 132$. This is not enough, so add 1 more bus:
$132 + 44 = 176$. Add 1 more bus again (to make 5 buses):
$176 + 44 = 176 + 40 + 4 = 180 + 40 = 220$
The answer is 5 buses.

Sample solution C:

$$44 \overline{)183}$$
$$\underline{132} \quad 3$$
$$51$$
$$\underline{44} \quad 1$$
$$7 \quad \text{We still need 1 more bus for the remaining 7 people.}$$
$$\underline{1}$$
$$5 \text{ buses}$$

Follow-Up and Discussion

Sample solution A illustrates the traditional method of long division, which many adults were taught is the only way to divide. The student performing the calculation has carried out the calculation properly, but does not appear to understand what the method is telling them. The remainder of 7 students won't have a place on a bus. The student who completed Sample solution B is using a self-generated method, and demonstrates an understanding of the situation. In Sample solution C, the student is correctly using an alternate algorithm or method called *repeated subtraction*, the conceptual basis of which is similar to the thinking shown in Sample B. The repeated subtraction algorithm is efficient for use with whole numbers, and is more conceptual in nature than the traditional long division procedure. Next, we will use manipulatives to explore the reasoning underlying the traditional procedure, and the repeated subtraction method.

Exploration/Task

To model the division statement $340 \div 14$, start by representing 340 using base ten blocks. Next, calculate $340 \div 14$ using long division, and then once again using repeated subtraction as illustrated in Sample solution C above. (For example, you might begin by subtracting 10 groups of 14 from 340.) As you perform each numeric step in each calculation, use the base ten blocks to illustrate what is happening. Which method makes more sense? Which one was easier to model with base ten blocks?

> **MATHEMATICAL TERM**
>
> A *quotient* is the result of dividing one number by another. The *remainder* is what is left when we divide to the nearest whole number quotient.

Follow-Up and Discussion

You may have found yourself struggling to model the division problem using long division, but that it was easier to do so using the repeated subtraction method. One of the problems with long division is that often only the leading digits of a number are recorded. For example when dividing 340 by 14, students used to be taught to say, "14 'goes in' to 34 twice." Using repeated subtraction, we would think, "There are at least 20 groups of 14 in 340," or, if we aren't sure, we can start by subtracting 10 groups of 14. Students typically find the repeated subtraction method easier to model and understand, and it is quite computationally efficient, at least for whole number division.

Sometimes the remainder is not paid much heed in whole number division. In Sample solution A for the school bus problem, the remainder of 7 was not taken into account when determining the number of buses. In numeric division problems, attention is often given to the answer—or *quotient*—but not to the *remainder*. Sometimes in a context, however, the remainder is important.

Exploration/Task

Each of the following story problems relates in some way to the calculation 23 ÷ 2, but the various contexts in which the calculation is presented lead to a different answer for each problem. Answer each problem before going on, and think about how the context influences the answer.

1. 23 pizzas were shared equally between 2 classes. How many pizzas did each class get?
2. A group of 23 small children was crossing a busy street with their Grade 8 helpers. Each helper was able to hold the hands of, at most, 2 small children. How many helpers were needed for the group to cross the street together?
3. There were 23 slices of bread for sandwiches. If 2 slices were used for each sandwich, how many complete sandwiches could be made?
4. There were 23 matching socks on the clothesline. Mom put them away in pairs and put the leftovers in the lost-socks bag. How many socks did she add to the bag?

Follow-Up and Discussion

You might have noticed that as you solved each problem, the role of the remainder was different. The remainder might have

1. added precision—the last pizza was split into 2 halves
2. been important—1 helper was needed to hold the hand of the remaining child
3. not been that important—only the complete sandwiches were being counted
4. represented the answer to the problem—in this case, what was left when the socks were matched

While it might make sense to have half a pizza, there are other contexts in which a less-than-whole number doesn't make sense—for example, when dividing children into equal groups. Again, the role of context is crucial in the interpretation of division, as well as connecting concepts in division of whole numbers, fractions, and decimals.

5.6 Factors, Multiples, Divisibility, and Prime Numbers

The term *multiple* can be remembered as it connects directly to the word *multiply*. For example, 10, 15, 20, 25, 100, and even 5010 are multiples of 5; this means that they are equal to 5 multiplied by some (whole) number. Similarly, 5 is a *factor* of 10, 15, 20, 25, 100, and even 5010, which means it divides evenly (or exactly) into each number listed.

The concepts of factors and multiples are often taught as numeric properties, separate from contexts, and thus without meaning. As a result, many students do not understand these ideas well or see their purpose. As teachers, we can help students see the usefulness of an idea by helping them connect it to other mathematical ideas and processes.

Understanding factors and multiples, for example, can help students grasp calculations with fractions, but teachers do not always explicitly refer to this connection when teaching these concepts. A flexible understanding of multiplication and division, combined with (or connected to) an understanding of the concepts of factors and multiples, can help students understand and even invent strategies that are useful when comparing, or calculating with, fractions. This section introduces the ideas of factors and multiples, and these ideas are further explored in chapters 6 and 7.

> **MATHEMATICAL TERM**
> A *multiple* of a number is that number multiplied by any other whole number.

> **MATHEMATICAL TERM**
> A *factor* of a number is a whole number that divides into that number with no remainder. (Proper factors exclude 1 and the number itself.)

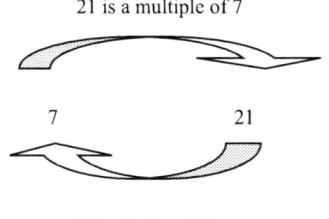

Here, 7 is shown as a factor of 21. Since 7 × 3 = 21, 3 is also a factor. In other words, both 7 and 3 divide into 21 with no remainder. And 21 is a multiple of both 3 and 7.

Some numbers have many factors. For example, 24 is divisible by 2, 3, 4, 6, 8, and 12—each of these numbers is a factor of 24. There are a number of combinations of numbers that can be multiplied to 24, for example, 3 × 8 or 2 × 3 × 4. Another way to think about divisibility is to consider the connection between multiplication and division. For example, if 3 × 4 = 12, we can see (possibly even by using a model, such as an area model, that has been used to solve a multiplication problem) that it is also true that 12 ÷ 3 = 4 and that 12 ÷ 4 = 3. We can say that 12 is *divisible* by both 3 and 4.

> **CONNECTION**
> For more on whole number division and what the division operation means, see section 5.3. To review different division models, see section 5.5.

Our previous investigation of area models has shown that we can use a rectangle or area model, or a similar model called an array, to illustrate the factor-product relationship:

One way to visualize the idea that 12 is divisible by 3 (or 4 or 2 or 6) is to show 12 as a rectangle with 3 (or 4 or 2 or 6) as one of the dimensions. Similarly, we can see that 12 is divisible by both 2 and 6 because we can make a rectangle with area 12, and side lengths of 2 and 6:

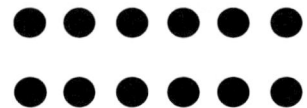

Exploration/Task

What about the factors of the number 11? What rectangles can you form using exactly 11 chips (also called circular counters)?

Follow-Up and Discussion

We find that no matter how hard we try, the only rectangle that we can form using 11 chips is a 1-by-11 rectangular shape—which looks like one long row of chips:

Since this long row is the only possible rectangular array with exactly 11 units, we know that 11 is divisible by only the whole numbers 1 and 11. The number 1 is called a unit, and 11 is the number we were investigating. The number 11 is called *prime*, meaning it has no factors other than 1 (the unit) and itself.

> **MATHEMATICAL TERM**
>
> A *prime number* is a whole number that is divisible by only the whole numbers 1 (the unit) and itself.

> **MATHEMATICAL TERM**
>
> A *composite number* is divisible by numbers other than 1 and itself.

Reasoning

The best way to understand the idea of prime numbers is to use visual reasoning as before: the only rectangle that can be formed with a prime number of chips uses all of them in a single row. More formally, a prime number is only *divisible* by 1 and the original number. Numbers that are not prime are called *composite* numbers.

But what about the number 1? The only way to represent 1 is with a 1-by-1 rectangle or 1-by-1 array.

●

Is 1 a prime or composite number? The number 1 is a special case, and mathematicians have disagreed about how to classify it. The number 1 is really neither prime nor composite, thus it is referred to as the *unit* in whole numbers. (Other sets of numbers may have other units).

Practice and Further Exploration

Use chips to explore the following values to determine if they are prime or composite. Note that finding even one rectangle with dimensions other than 1 and the number is enough to show that the number is composite. As you work, think about how many numbers you need to test before determining the first side-length.

a) 10
b) 13
c) 18

Follow-Up and Discussion

While 13 chips form only a 1-by-13 array, you may have found a number of models to illustrate both 10 and 18. For example, 18 can be a 2-by-9 or 6-by-3 rectangle. Notice that

$$2 \times 9 = 18 \text{ and } 6 \times 3 = 18$$

2 and 3 are prime numbers, and 9 and 6 are not. We can further factor both expressions:

$$2 \times 9 = 18 \text{ can be written } 2 \times 3 \times 3 = 18$$

and similarly,

$$6 \times 3 = 18 \text{ can also be written } 2 \times 3 \times 3 = 18.$$

The factorization $2 \times 3 \times 3$ is called the *prime factorization* of 18; in plain language this means that none of the numbers multiplied in the expression can be factored any further because all of the numbers are prime.

In contrast, if we explore the factorization of 100, we find that 10 is a factor of 100, but it is not a prime factor. 10 is not a prime number as it can be further factored into 2×5. The prime factors of 100 are 2 and 5, and the *prime factorization* of 100 is $2 \times 2 \times 5 \times 5$. The factors of 100 are 2, 4, 5, 10, 20, 25, and 50, but the prime factors of *these* numbers are all 2 and 5.

What if we want to determine if a very large number—too large to be conveniently represented with chips—is prime or composite? How might we quickly determine if a large number, say 323, is prime or composite?

> **MATHEMATICAL TERM**
>
> A *prime factor* has no factors other than 1 and itself.

Exploration/Task

Divide 323 by a few numbers, looking for factors. How many values would you need to test to determine whether it is prime or not? Why?

Follow-Up and Discussion

If you tried enough numbers, you may have found the factors 17 and 19, since $17 \times 19 = 323$. We know that finding any factors of 323 other than 1 and 323 is enough to determine that 323 is not a prime number. However, if 323 had not been divisible by 17, could you have drawn the conclusion that it was a prime number? Do we need to test all possible values from 2 to the number itself to determine if it is prime?

Example

If we want to test a number, for example 437, to determine if it is prime, we need to know if it is necessary to try to divide it by every whole number up to 437. If not, how far is necessary? The calculations to follow explore and explain why it is necessary to test various values. (Can you predict in advance the largest number you will need to test? Hint: 20×20 is 400, and 21×21 is 441. If one factor is more than 21, will the other be more or less than 21? What does this say about the values we need to test?)

Possible value	Do we need to test if it is a factor?
2	437 is not an even number, so 2 will not be a factor. If 437 were even, we could conclude right away that it isn't prime.
3	Need to test it—but 3 does not divide evenly into 437.
4	Since 437 is not even, and is not divisible by 2, it is not divisible by 4 either.
5	437 does not end in 5 or 0, so is not divisible by 5.
6	We don't need to test to know that it is not divisible by 6; a number that is divisible by 6 would also be divisible by 2 and 3.
7	We need to test this one—but the test determines that it is not a factor.
8	The number would have to be even to be divisible by 8.

At this point, it might be apparent that if a test value such as 6 or 8 has factor values that we have already tried, then we don't need to test it. In other words, if a number divides by 6 (meaning 6 is a factor), then it will also have 2 and 3 as factors, which we have already tried; we can conclude that we really only need to test the *prime* numbers to see if they are factors. Composite numbers will have factors less than the number, so we would have already tried them. We can continue to test only prime numbers to determine if they are factors:

11	Not a factor
13	Not a factor
17	Not a factor
19	437 is, in fact, divisible by 19. We find that $437 = 19 \times 23$.

Follow-Up and Discussion

Since we have found that 19 divides into 437, we can conclude that 437 is not prime. But, if 19 had not divided into 437, how would we have continued with our test? We would not need to test 20 and 21 as they are not prime, and so we have already checked the factors of these numbers for divisibility into 437. But would we need to continue past 21?

Reasoning

The key observation in this case is that the square root of 437 is between 20 and 21. The *square root* of a number is the number that is multiplied by itself to result in the original value. We know this because 20 × 20 is less than 437, and 21 × 21 is more than 437. Using similar reasoning, if a factor of 437 were greater than 21 (such as 23), then the *other* factor in the pair would have to be less than 21 (such as 19), to keep the product from being too high. Because we were working through the numbers in order, we would already have checked the smaller numbers, so we would already have found the smaller factor if there were one. We can conclude that testing prime numbers *up to the whole number just larger than its square root* is sufficient to conclude if a number is prime or composite.

> **MATHEMATICAL TERM**
>
> The *square root* of a number is the number that is multiplied by itself to result in the original value. For example, 20 is the square root of 400 because 20 × 20 is 400.

Practice and Further Exploration

Determine if the following numbers are prime or composite using a method of your choice:

a) 117
b) 1001
c) 199

Example

There is often a benefit to computation if we recognize divisibility, and are able to apply the associative property. For example, to find 16 × 34 ÷ 17 we might notice that 34 = 2 × 17:

$$16 \times 34 \div 17 = 16 \times \underbrace{2 \times 17}_{\text{34 was factored to } 2 \times 17} \div 17 = 16 \times 2 \times \underbrace{17 \div 17}_{\text{We know } 17 \div 17 = 1} = 16 \times 2 \times 1 = 32$$

Follow-Up and Discussion

To perform the calculation, we have used the concept of *inverses* from section 5.4, as well as the idea that multiplication and division can be grouped and performed in any order (the associative property). We also represented 34 as a *multiple* of 17, which might also have been recognized by knowing that 17 is a *factor* of 34 (34 = 17 × 2). Rewriting 34 as 2 × 17 makes the calculation much simpler.

Presenting students with numeric problems, and encouraging them to look for clues and shortcuts to solve them, is one method to motivate interest in factors and multiples. This will also help them to develop flexible numeric strategies that make sense.

Practice and Further Exploration

Look for ways to calculate each problem, taking advantage of the ideas of factors and multiples when possible. Remember that multiplication is commutative, but division is not. To solve the problems you may need to expand your notion of the order of operations. This will be further discussed in section 5.7.

Problem A:
$$2 \times 5 \times 12 \div 24$$

Problem B:
$$8 \times 33 \div 22 \div 12$$

5.7 Rethinking BEDMAS

BEDMAS is an acronym that stands for "brackets, exponents, division, multiplication, addition, and subtraction." In many classrooms, it is used so often that some students think it is, in fact, a mathematical term, rather than an acronym used to remember the order of operations. Memorizing this rule without truly understanding it precludes a lot of the flexible computational strategies that have been explored throughout the book. Rather than memorizing the acronym, we recommend that students *understand* the order of operations. Put simply, *order of operations* means that, first, all computations in brackets (parentheses) are addressed, then any multiplication or division (exponents actually imply multiplication; this will be explored in greater detail in section 12.9), and then addition and subtraction.

Using Problem B from section 5.6, we can make the following calculation. Note that the techniques used to complete the calculation involve many of the concepts explored throughout chapters 4 and 5, including the commutative property, the associative property, inverse operations, and factoring.

$$8 \times 33 \div 22 \div 12$$
$$= 2 \times 4 \times 3 \times 11 \div 22 \div 12$$ — Factor to try to simplify: We note that 22 and 33 both have 11 as a factor.
$$= 4 \times 3 \times 2 \times 11 \div 22 \div 12$$ — Seeing possible inverses, we also recall that multiplication is commutative.
$$= 4 \times 3 \times 22 \div 22 \div 12$$ — Multiplication and division are inverses.
$$= 12 \times 1 \div 12$$
$$= 12 \div 12$$ — More inverses!
$$= 1$$

Some students might feel the calculation shown above contradicts the BEDMAS rule. If you are unsure if the calculation is correct, perform the original calculation, working from left to right, using a calculator. You should find that you get the same answer.

The order of operations becomes more important when complex algebraic expressions are used. For example, if we did not all agree on the order of operations, then expressions such as $4y^2 + 3y - 8$ would have to be written as $((4 \times (y^2)) + (3 \times y) - 8$, which is much more cumbersome than the first way. Order of operations simply allows us to use fewer brackets when writing such expressions. A deeper understanding of order of operations can make it easier to perform many calculations, as we saw when we examined factors, inverses, and so on.

5.8 Fact Families

Given the availability of inexpensive calculators, as well as the emphasis on conceptual understanding and problem solving, some teachers are uncertain what role basic facts play in learning mathematics. The short answer is that basic facts are still very important. What is not important—what is, in fact, detrimental—is to ask students to demonstrate speed-related fluency of facts, particularly in stressful, timed situations such as tests or competitive games. Addition and subtraction of larger numbers is greatly assisted by knowing basic addition and subtraction facts up to 20. By making explicit connections between addition and subtraction, the number of facts that students need to know can be reduced. For example, knowing that $3 + 5 = 8$ helps them know that $8 - 5 = 3$.

Other strategies can and should be developed to assist students in doing quick mental calculations when they do not yet know facts by rote. For example, knowing that $9 + 10 = 19$ helps students find the answer to $9 + 9$, as it must be just 1 less. If students are encouraged to attend to and share their reasoning, they will naturally develop multiple strategies for problem solving and fluency will follow. Providing multiple opportunities for students to participate in engaging activities and games supports the internalization of rote facts without stress and frustration. For example, a play store with pretend currency is an engaging context that supports the development of fluency in mental adding and subtracting. Many other games and web-based activities are available that have similar positive effects.

It is generally accepted that students should learn their multiplication facts up to 10 or 12. Explicitly connecting multiplication facts to division is also important. For example, the facts $6 \times 7 = 42$ and $7 \times 6 = 42$ should be simultaneously connected to $42 \div 6 = 7$ and $42 \div 7 = 6$. This group of four facts really contains just one main idea or relationship, as embodied in the area model below. These related facts—called fact families—should always be explicitly connected.

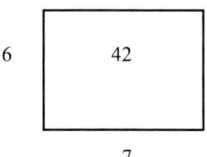

Although games can also be used to support increased fluency and, ultimately, the rote learning of multiplication facts, some organizational structure can be helpful. It is generally suggested that carefully ordering and grouping the facts is the best approach. For example, most students find that the 1 and 2 times tables are not difficult; the idea of doubling numbers comes quickly because of its relationship to addition facts. The fives and tens can be grouped together. The 4 times facts might be approached using *double twos*, which also will yield the eights (using *double double*).

	1	2	3	4	5	6	7	8	9	10
1	1	2	3	4	5	6	7	8	9	10
2	2	4	6	8	10	12	14	16	18	20
3	3	6	9	12	15	18	21	24	27	30
4	4	8	12	16	20	24	28	32	36	40
5	5	10	15	20	25	30	35	40	45	50
6	6	12	18	24	30	36	42	48	54	60
7	7	14	21	28	35	42	49	56	63	70
8	8	16	24	32	40	48	56	64	72	80
9	9	18	27	36	45	54	63	72	81	90
10	10	20	30	40	50	60	70	80	90	100

To keep track of the multiplication facts that students have mastered, and those that remain to be learned, a 10-by-10 grid can be used to show facts up to 10 × 10. Certainly, without some order and strategy, the number of facts (10 × 10 or 100) looks daunting.

Thinking of the facts as pairs based on the commutative property, for example, knowing that 3 × 2 = 2 × 3, immediately reduces the number of facts to about half. Simple facts such as those containing a factor of 1 or 10 are also reasonably easy for most students to learn. Next come the twos (doubles) and fives; again, many children easily learn these through games and activities. Mastering these facts reduces the number of facts that students must remember. Crossing out the ones, twos, fives, tens, and one set of the paired facts greatly reduces the number of facts that students must still learn.

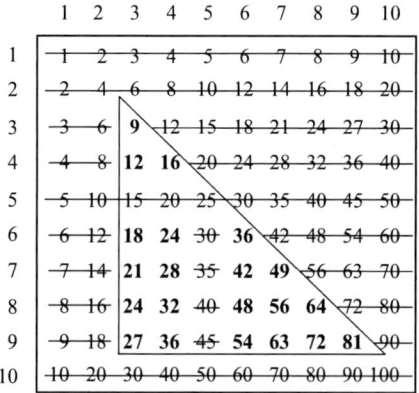

The remaining facts, those in bold type inside the triangular outline, can then be learned in groups. Three times facts are often learned quite quickly, for example, using doubles plus 1 more. Four times facts can be learned by doubling the twos (which are already familiar), and eights often follow reasonably easily by doubling the fours. Nine times facts can be fairly easily calculated by knowing the 10 times facts and subtracting—for example, 9 × 6 = 10 × 6 − 6. Using these ideas, learning the remaining facts becomes much more manageable.

Skip counting, for example, by 2 (2, 4, 6, 8...) or 5 (5, 10, 15, 20...), in the early years can also prepare students to learn these basic facts. For example, skip counting by 5 relates to 5 × 1, 5 × 2, 5 × 3, and so on.

Array models (see section 5.6) are yet another method that can help students relate, connect, and calculate facts. For example, modelling 8 × 6 with an array using counters encourages the use of strategies. Visualizing the 8 × 6 array as two 4 × 6 or 8 × 3 arrays makes it clear that 8 × 6 is two groups of 4 × 6 or 8 × 3; therefore, the result must be 2 × 24 or 48.

Exploration/Task

Using counters, model 8 × 6, and then rearrange your model to show rectangles with dimensions 4 × 12 and 2 × 24 using the same number of counters. What does this demonstrate?

Follow-Up and Discussion

The use of arrays encourages flexible understanding of numbers and operations, and can help students develop strategies for multiplication facts. The halving and doubling strategy that was used to change 8 × 6 to 4 × 12, and then to 2 × 24—which can be done fairly easily mentally—is a useful method for other calculations. For example, using a doubling strategy, 4 × 6 is 4 × 3 doubled, and 4 × 8 is 4 × 4 doubled. We can also add on to a known fact, for example, 6 × 8 is 5 × 8 + 8. Again, students should be encouraged to develop, use, and share these strategies. Flexible understanding of numbers and operations at an early age can greatly help students to develop and use such methods. The ultimate development of rote fact recall does speed up much of the operational aspect of problem solving. Rote recall of facts can be faster than using a calculator and is, obviously, not prone to keystroke error.

In summary, students should be encouraged to develop and use their own strategies for determining unknown facts until they are known by rote. A purposeful and logical learning order of facts greatly reduces the number of facts to be remembered. Strategies are important to develop also for facts that are not yet learned by rote.

Strategies for developing fluent use of methods such as those described above, and, ultimately, rote recall, are not only useful when initially introducing operative facts to students, but also make excellent interventions for students who are struggling. Students can become frustrated if they cannot remember facts that involve larger values, such as 8 × 7 or 9 × 6. We suggest, rather than using flash cards and other memorization techniques, continued emphasis on strategies that connect unknown facts to those students already know. Co-operative games and other activities that motivate and interest students can support fluency without the emotional stress that accompanies competitive activities or timed tests.

Chapter Problems

1. Choose two different, single-digit numbers. Use these numbers as the dimensions of an area model as could be used demonstrate the commutative property of multiplication. Use the same area model to demonstrate that multiplication can be thought of as repeated addition.
2. In a Grade 3 class there are 17 students. Each student has 12 coloured pencils. Sketch how a student might use base ten blocks, or a drawing of base ten blocks, to model the process of finding the total number of coloured pencils in the class.
3. Consider the following multiplication statement: 2 × 18 × 5 = ?. How might a student use the associative property of multiplication to make the calculation friendlier? Extend this thinking to the problem 4 × 37 × 5 ÷ 2 = ?.

4. A class of Grade 4 students has 144 cookies to sell. They are selling them in packages of 6. Your students use a repeated subtraction method to determine how many packages they can prepare with the 144 cookies. Calculate the answer using this method in two different ways; that is, by subtracting different numbers of groups of 6 each time. One student begins by subtracting 10 groups of 6, while a second begins by subtracting 20 groups of 6. Complete each solution method, and compare them. (Your two examples might illustrate the work of students who are relatively less and more computationally fluent.)
5. Your students are struggling with the following multiplication facts: 8×8, 8×9, and 9×7. Other than memorizing the facts, what strategies might help the students learn these facts?

Further Reading

Buczynski, S., Gorsky, J., McGrath, L., and Myers, P. (2011). Sift like Eratosthenes. *Teaching Children Mathematics, 18*(2), 110–118.

> This article details a Grade 5 exploration of prime numbers using the sieve of Eratosthenes. The lesson is designed to increase student understanding of prime numbers through exploration.

Kurz, T. L., and Garcia, J. (2012). Moving beyond factor trees. *Mathematics Teaching in the Middle School, 18*(1), 52–60.

> Kurz and Garcia discuss the use of factor trees and how they are not based on a conceptual understanding of numbers. They provide ideas for different activities that can be used in a classroom, including resource sheets, to deepen student understanding of divisibility, as well as greatest common factors and least common multiples.

Lamberg, T., and Wiest, L. R. (2012). Conceptualizing division with remainders. *Teaching Children Mathematics, 18*(7), 426–433.

> Lamberg and Wiest discuss concrete tasks used with Grade 3 and 4 students to help them to conceptually develop division with remainders. They discuss using the tasks to move from a concrete representation of division to more formal operations built on students' understanding.

Lee, J. K., Choi, K. M., and McAninch, M. (2012). An exce-L-ent algorithm for factors and multiples. *Teaching Mathematics in the Middle School, 18*(4), 236–243.

> The authors combine Singaporean and Korean procedures to help students learn to identify least common multiples and greatest common factors. The article details the procedure they used and shows how it links to the factor trees that are often used in schools.

O'Connell, S., and SanGiovanni, J. (2011). *Mastering the basic math facts in multiplication and division.* Portsmouth, NH: Heinemann.

> The authors break down the basic facts of operations for 0 to 10. They discuss strategies to help students learn and understand the facts for multiplication and division without relying on rote memorization.

Chapter 6

Fraction Representations and Additive Operations

6.1 Fractions: Concept and Contexts

Simple concepts of fractions are often embedded in everyday language. For example, most children know what "give your brother half the chocolate bar" means, yet many adults describe fraction concepts as something they have never understood well. In everyday life, many activities such as cooking, travelling, and telling time require the use of less than a whole unit of a quantity.

While computational fluency with very difficult fractions is no longer required to the degree that it was before calculators, learning fraction operations is still important. Many algebraic manipulations in higher mathematics, for example, require an understanding of *rational* expressions, which are algebraic representations that involve fraction concepts. Secondary students who do not know how to simplify an expression such as

$$\frac{5}{(x-1)} - \frac{4}{(x+1)}$$

are often stumped by the fraction concepts related to the expressions $(x - 1)$ and $(x + 1)$, rather than the other algebraic aspects of the problem. Another important reason for students to understand how to operate with fractions relates to the issue that some fractions simply do not convert easily or exactly to decimal numbers (more on this to come).

One reason why fractions may have traditionally proven so difficult for many people may be the early imposition of computational rules in school. Such procedural instructions are often not accompanied by sufficient concrete experience. As a result, many students struggle to develop skills without the related conceptual reasoning—and no wonder the rules leave many students confused! How nonsensical does a seemingly arbitrary rule like *invert and multiply* sound without understanding its basis.

58 Chapter 6

In this chapter, we will work extensively with concrete representations and models, and will illustrate how the symbolic algorithms are indeed necessary outcomes of our explorations with concrete models. Such algorithms will be explicitly derived from, or connected to, these models.

There are a number of uses and interpretations of the fraction notation. For example, we can interpret the notation $\frac{6}{3}$ as division, meaning "6 divided by 3," and obtain the answer of 2. Alternatively, the notation $\frac{6}{3}$ can sometimes be used to imply a ratio.

A fractional or *part-whole* interpretation of $\frac{6}{3}$ assigns a particular meaning to the value in each location in the fraction notation. Continuing with this interpretation, consider this definition:

> **CONNECTION**
>
> For a description of ratios as representing division or the comparison of whole numbers, see section 5.4.

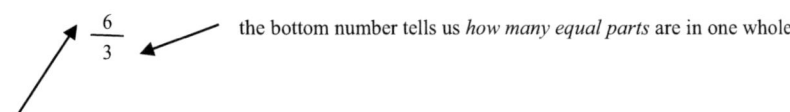

the bottom number tells us *how many equal parts* are in one whole

and the top number tells us *how many of these parts* we have.

Using this part-whole interpretation, the 3 in the (bottom of the) number tells us that one whole is divided into *3 equal parts*:

Thus the "unit" or size of the piece we are referring to in this case is $\frac{1}{3}$ (because there are 3 parts in 1 whole, so the size of each piece is $\frac{1}{3}$). Understanding *both* the implied one-whole unit and the newly defined "unit" of one-third are fundamentally important ideas.

For example, in a given fraction using thirds, each shape representing $\frac{1}{3}$ must be equal in net size. The shape illustrated to follow could be used to represent one of these units or parts of size $\frac{1}{3}$:

> **INSIGHT**
>
> One whole must be clearly defined for a fraction to make sense. A fractional unit, such as $\frac{1}{3}$, then becomes a new unit, but this new unit only makes sense *with respect to the original one whole*.

Continuing with the same fractional unit, $\frac{1}{3}$, let's examine a quantity of them. For example, using the example of $\frac{6}{3}$, the top number tells how many pieces we have, so the 6 in $\frac{6}{3}$ means we have 6 *of these pieces* (that is, 6 pieces of size $\frac{1}{3}$). Here is one possible model of $\frac{6}{3}$:

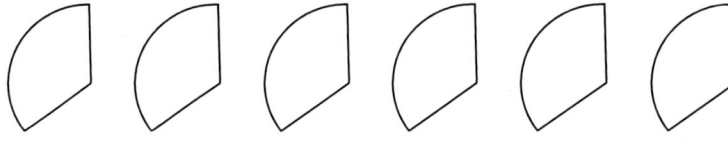

We can also group the pieces together into wholes, in which case it might be observed that $\frac{6}{3}$ is equivalent to 2 (which we understand to mean 2 wholes).

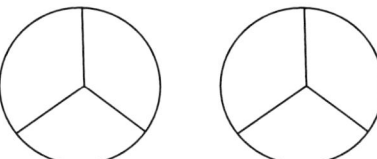

While we ultimately obtain the same numeric answer as with the division interpretation of $\frac{6}{3}$ (6 divided by 3 is 2), the model may look different. Sometimes sets or array models are used to represent fractions. For example, the illustration below shows an array model of $\frac{3}{8}$. The 8 small circles are taken to mean one whole.

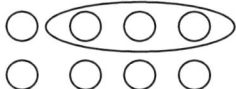

Fractions are used in a wide range of industrial and retail situations, in everything from tools such as $\frac{5}{8}$-inch wrenches, to $\frac{1}{4}$-metre lengths of fabric, to half-price sales. Making use of contexts is critically important if students are to understand fractions and be able to interpret them appropriately. Not only does context make the learning more interesting, useful, and real, it helps students visualize reasonable answers, and what they actually mean.

6.2 Exploring Representations, Models, and Manipulatives

Many students begin to study fractions with preconceived misconceptions, often as the result of the rules or procedures that well-meaning adults have taught them. For example, when studying whole numbers students may have learned a rule such as *you always have to divide by the smaller number*. Such rules are often incomplete (and thus inaccurate) and based solely on memorization. The use of models and representations, on the other hand, is necessary for developing conceptual understanding.

There really is no single model or manipulative that is best to use when working with fractions. Many schools use fraction circles, which are plastic disks cut into equal parts, usually up to twelfths, that allow students to explore the part-whole meaning of fractions. The first few in the set might look like this:

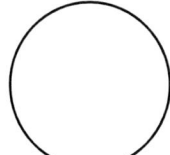 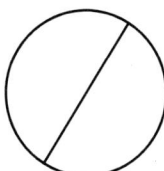 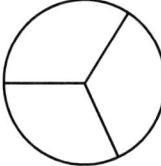 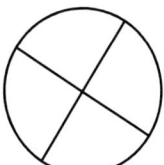

The advantage of using fraction circles is that it is very easy to identify one whole—a complete circular piece.

Another popular manipulative is the fraction bar or fraction strip. Teachers can purchase plastic fraction bars, or students can cut their own templates out of card stock. Some teachers feel that it is well worth the time to have children cut out the fraction pieces because of the understanding they develop in doing so. The other advantage of having students cut out their own templates is that they can be put in zip-lock bags and easily taken home. When the fraction strips are lined up in one-whole units, they might look like this (but the fractional pieces can also be cut apart). Again, it is very important that the one-whole bar be clearly specified.

Students may also choose to draw various different pictures to represent fractions, depending on the context. This is to be encouraged and their ideas should be shared. The greater the number of representations, the better the tool kit that children have available to solve a particular problem.

Another interpretation of fractions involves the use of sets, areas, or arrays. In these contexts, one whole is taken to mean an entire group or region. For example, the following might be defined as a given area:

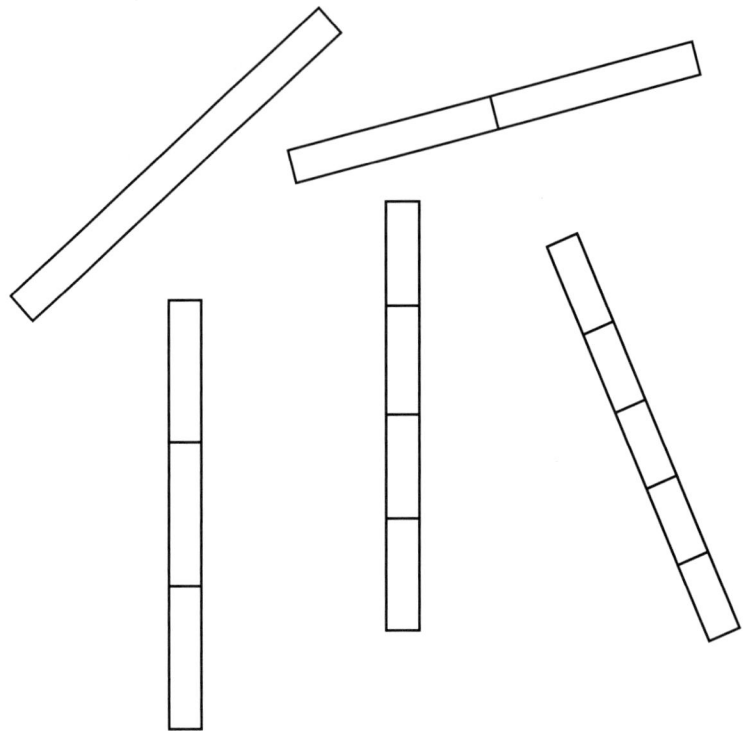

The regions A and B could be $\frac{1}{8}$ and $\frac{1}{4}$ of the total area respectively. Alternately, consider the following array, which is also taken to represent a whole:

Circling two of the dots might be written as either $\frac{2}{16}$ or $\frac{1}{8}$. Discussing these different number-names helps students develop an initial understanding of equivalence.

A very important aspect of fractions learning is the investigation of real-world problems that can be answered using models and reasoning, rather than by initially emphasizing formal numeric or symbolic methods. Problems related to each of the four fundamental operations should be explored before any connections to formal numeric methods are made in order to develop understanding and reasoning.

Exploration/Task

Using manipulatives or hand-drawn models as you choose, use reasoning (rather than formal procedures) to answer the following problems:

1. Two cakes of the same size were partially eaten. There was one-third of one cake left, and half of the other. How much cake was left in all? Clearly state what you are defining to be one whole, and express the remaining cake as a single fraction value.
2. There are $2\frac{1}{2}$ metres of chord on a spool. If you purchase $\frac{3}{4}$ of a metre, how much is left on the spool?
3. It takes $\frac{1}{2}$ litre of paint to cover one door, and there are $1\frac{1}{2}$ litres of paint in a can. How many doors can be painted using this can?
4. There is $\frac{3}{4}$ of a pizza left in the box. If you ate half of the remaining pizza, how much of the whole pizza would you have eaten? Try to solve this problem more than one way.
5. Pieces of ribbon are being cut for a school craft. There are 3 metres of red ribbon. If $\frac{2}{3}$ of a metre is needed for each craft, how many crafts can be made using red ribbon?

Compare your solutions with a colleague if possible and share your models. Discuss the operation you were modelling. Is there more than one way to answer each question?

Follow-Up and Discussion

In fact the above problems span all of the fundamental operations—addition, subtraction, multiplication, and division. And yet, each could be answered using reasoning and models, and often using more than one approach. For example, for Problem 1, we might represent the $\frac{1}{2}$ and $\frac{1}{3}$ with manipulatives as sixths, using trial and error, and then add the sixths together. Alternately, combining the $\frac{1}{2}$ and the $\frac{1}{3}$ and comparing them to one whole shows that the answer is $\frac{1}{6}$ *less* than one whole, or $\frac{5}{6}$. It is possible to use similar strategies to solve each of the problems.

6.3 Comparing Fractions

In this section, we will explore how we can use models to compare fractions, without using formal procedures. If children do not work with concrete materials, it might seem to them at first glance that a fraction such as $\frac{2}{12}$ is bigger than $\frac{1}{5}$ because the numbers seem bigger in $\frac{2}{12}$; however, there are two parts to the fraction notation, and it is the bottom number, called the *denominator*, that determines the *size* of the fraction unit. Extensive hands-on experience is needed to become comfortable with the idea that a *larger* denominator means that the size of the unit is *smaller* (since more pieces are needed to make up one whole).

Working with concrete manipulatives helps students develop the ability to visualize basic fraction sizes. After working with fraction strips or circles, we can visualize, for example, that $\frac{3}{4}$ is more than $\frac{1}{2}$. A helpful model to introduce (if it does not emerge on its own) is the open number line. The number line is often used to show the relative sizes of numbers. In an open number line, we do not necessarily have to start at 0 and put in every single value, rather we place the values as needed in their relative positions. We can also put benchmark fractions, which are easy numbers used for comparison, on the number line.

> **MATHEMATICAL TERM**
>
> The *denominator* is the bottom number in the symbolic notation of a fraction, or the number of pieces that the whole has been divided into.

Example

If a student is asked to compare $\frac{3}{4}$ and $\frac{3}{8}$, she might draw the following:

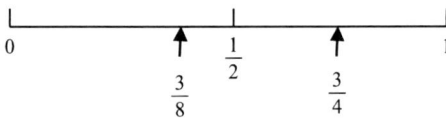

Follow-Up and Discussion

Although it is difficult to identify her reasoning from the model alone, the child might have used this model to compare $\frac{3}{8}$ and $\frac{3}{4}$ by comparing them *both* to $\frac{1}{2}$. She might reason that $\frac{3}{8}$ is smaller than $\frac{4}{8}$, so it must be smaller than $\frac{1}{2}$. She might also know that $\frac{3}{4}$ is larger than $\frac{2}{4}$, which is the same as $\frac{1}{2}$, and so could deduce that $\frac{3}{8}$ is smaller than $\frac{3}{4}$.

Another method for comparing fractions, which is used less frequently yet can be very helpful, is comparing denominators. Continuing with the comparison of $\frac{3}{8}$ and $\frac{3}{4}$, we can see that there are the *same* number of pieces in $\frac{3}{8}$ and $\frac{3}{4}$. However, the $\frac{1}{8}$ pieces are smaller than the $\frac{1}{4}$ pieces, so $\frac{3}{8}$ must be smaller.

Hands-on materials should initially be used to help students make comparisons. It may take some time for them to accept that $\frac{1}{2}$ and $\frac{2}{4}$ represent the same quantity; indeed, there is a mathematical abstraction at work here. After all, $\frac{1}{2}$ is represented by *one piece* while $\frac{2}{4}$ contains *two pieces*

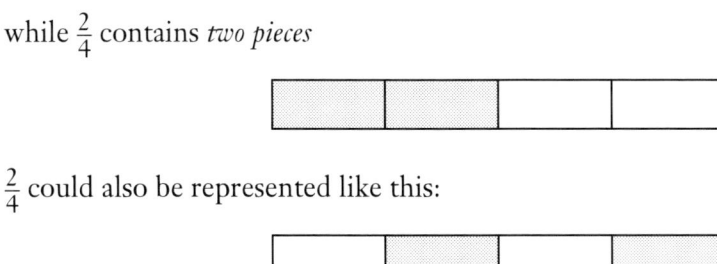

$\frac{2}{4}$ could also be represented like this:

After sufficient experience, students will generalize that it is the overall space or length used—rather than the number of sections—that indicates the size of a fraction. Sometimes it is easy for adult learners to forget that such concepts, which might seem obvious, do in fact require learners to generalize and develop higher levels of abstraction.

Children tend to intuitively realize that if they are comparing fractions of the same size (with the same denominator), it is the top number—or the *numerator*—that matters. The numerator indicates how many pieces we have, and if the pieces are the same size (if the denominators, or bottom numbers, are the same), then we can simply compare numerators. Alternately, if the numerators are the same, we know we have the same total number of pieces, so now it is the *size* of the pieces—the denominator—that matters. A smaller denominator represents larger pieces, which is a challenging idea unless it is conceptually understood.

> **MATHEMATICAL TERM**
>
> The *numerator* is the top number in the symbolic notation of a fraction, or the number of pieces that we have.

Exploration/Task

Compare the fractions $\frac{3}{11}$ and $\frac{3}{12}$ using reasoning.

Follow-Up and Discussion

In thinking about this task, you might have realized that even though there are the same number of *pieces* in each fraction, the twelfths are *smaller* pieces than the elevenths. Thus the same number of smaller pieces must be less in total; therefore, $\frac{3}{12}$ is less than $\frac{3}{11}$.

Given enough time and the appropriate materials, intuitive methods will naturally emerge. For example, comparing $\frac{2}{12}$ and $\frac{1}{5}$ might be done in two stages: first, by using concrete materials to recognize that $\frac{2}{12}$ and $\frac{1}{6}$ are equivalent, and then by comparing the denominators of $\frac{1}{6}$ and $\frac{1}{5}$. Another method might be to notice that $\frac{1}{5}$ and $\frac{2}{10}$ are equivalent, and then compare the tenths and twelfths.

In the past, a great deal of attention was given to a single strategy for comparing fractions. This strategy involves the formal numeric method of converting both or all fractions to the same denominator. While this method works all of the time, there are many cases (as we have just seen) that do not require the use of this formal method, and are, in fact, easier to resolve in other ways. This *common denominators* method relies on comparing numerators after any or all fractions have been converted to equivalent fractions with the same denominator.

> **MATHEMATICAL TERM**
>
> Fractions that have the same denominator are said to have *common denominators*.

> **CONNECTION**
>
> For more formal numeric methods of finding common denominators, see section 6.6.

Practice and Further Exploration

Compare each of the following fractions using (informal) modelling and reasoning strategies similar to those described in this section. Note your strategy as you work.

1. $\frac{2}{3}$ and $\frac{1}{2}$
2. $\frac{13}{12}$ and $\frac{12}{11}$
3. $\frac{4}{6}$ and $\frac{3}{5}$

6.4 Concrete Equivalent-Fraction Models

> **INSIGHT**
> Fractions need to be represented in the same size unit before adding or subtracting.

Section 6.3 explored a number of methods for comparing fractions without using formal numeric procedures, which lays important groundwork for understanding compatible units when adding and subtracting fractions. After sufficient hands-on exploration, students will already informally recognize that sixths work as a unit when adding or subtracting halves, thirds, and/or sixths, and that tenths work for halves and fifths. Similarly, halves, fourths (quarters), and eighths have a relationship. It might take a slightly deeper understanding and a little more experience to recognize that adding thirds and fourths requires twelfths.

When using manipulative pieces to find equivalent fractions, a student's first strategy might involve guess-and-check. As students gain experience with manipulatives, their representations will become more intentional.

Early conjectures that students make about the correct unit form important foundations that lead to the formal construction of the *common denominator*. For example, they might conjecture that to find a fraction unit that allows them to add two given fractions, they need to multiply the bottom numbers together. For example, to add $\frac{1}{3}$ and $\frac{1}{4}$, both fractions might be represented as twelfths—the product of 3×4. After exploring a number of similar models, the relationship of the 12 to the 3 and the 4 in the model might become apparent.

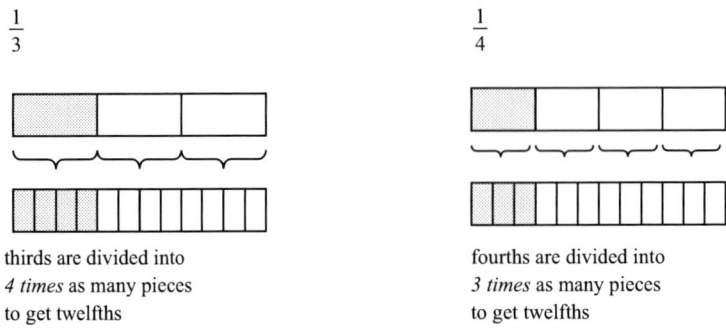

thirds are divided into
4 times as many pieces
to get twelfths

fourths are divided into
3 times as many pieces
to get twelfths

It seems we can multiply the numbers in the denominators of two given fractions to find a fractional unit that can represent both fractions. We might ask, is this always the case? Why does it work, and is there an easier method?

As students develop such conjectures, the ideas might be recorded in the classroom and refined as explorations progress. For example, it is fairly easy to show that the initial "multiply the bottom numbers" conjecture always works, but may not always result in the simplest or most reasonable unit. For example, a refinement might be that *if one number divides evenly into the other, you can just use the larger number (which is the smaller fraction piece) for the unit.* Eventually, students begin to understand that, sometimes, even if one number does not divide into the other, a smaller unit than the product of the two denominators might work. This is the case, for example, when adding fourths and sixths. While we *can* multiply 4 × 6 to get 24, and use twenty-fourths as the new denominator, this is not necessarily the easiest method. Experience shows that when adding fourths and sixths, we can also represent the values using twelfths rather than twenty-fourths.

6.5 Concrete Models and Strategies for Adding and Subtracting Fractions

As is the case when comparing fractions, a surprising number of fractions can be added or subtracted using only manipulatives, models, and reasoning, without the introduction of formal numeric or symbolic procedures. Concrete exploration develops reasoning, estimation concepts, an intuitive sense of fraction size, and initial experience with useful methods.

As is typical of problem-solving contexts, methods developed by children may be as varied as the children themselves. The problems to follow can be solved using a variety of methods. On a pedagogical note, when structuring classroom problems it is helpful to carefully monitor the progression of ideas. For example, problems that can be *directly* done with manipulatives (such as $\frac{4}{6} + \frac{1}{6}$) are easier than those that require the use of equivalency (see section 6.4), such as $\frac{1}{5} + \frac{3}{10}$.

Exploration/Task

Explore the following problems. If possible, complete each problem using more than one method. Use manipulatives and models, rather than any previously learned numeric procedures. Work with a colleague if possible and share your ideas. Remember too that if you were offering problems to your students, it would be best to frame each one in a context. As you work, you might think about an appropriate context for each; for example, a simple context for Problem 1 might involve two 8-slice pizzas, one with 3 slices left, and one with 2 slices left.

1. $\frac{3}{8} + \frac{2}{8}$
2. $\frac{1}{2} + \frac{1}{3}$
3. $\frac{2}{3} + \frac{3}{6}$
4. $\frac{3}{5} - \frac{1}{5}$
5. $\frac{3}{5} - \frac{3}{10}$
6. $\frac{5}{4} - \frac{3}{8}$
7. $\frac{7}{5} - \frac{1}{2}$

Follow-Up and Discussion

As noted, many of the above problems can be solved in several ways. Provided are a few approaches that could have been used to solve them. These approaches are meant to be illustrative, as other models are possible in many cases.

1. This question is a relatively straightforward initial question as the fractional units are the same. Using $\frac{1}{8}$-size pieces, and combining 3 of the pieces with 2 of the pieces, shows that $\frac{5}{8}$ is the answer. Many common contexts, such as the pizza scenario suggested earlier, are appropriate.

2. The units in $\frac{1}{2} + \frac{1}{3}$ are not the same, as can be seen when the sum is modelled.

   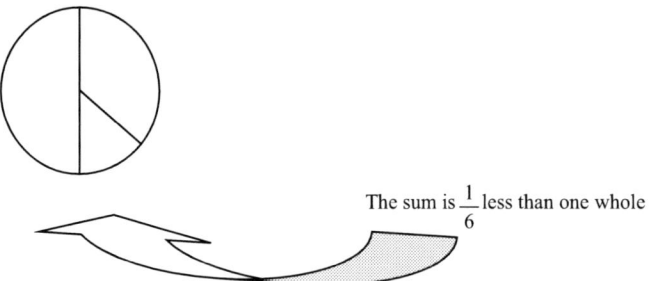

   The sum is $\frac{1}{6}$ less than one whole

   There are two typical approaches to problems of this type. The direct approach involves exploring manipulative pieces until a size that fits both fractions is found, in this case, sixths. Once both parts are represented as sixths, they can be added. Another approach that works well with concrete materials is to explore the size of the piece that is missing from—or the amount *less* than—one whole. Once the missing piece is identified as $\frac{1}{6}$, it can be reasoned that the sum is the rest of the whole, or $\frac{5}{6}$.

3. When using manipulatives to represent both fractions in the problem $\frac{2}{3} + \frac{3}{6}$, it might be tempting to re-represent $\frac{3}{6}$ as $\frac{1}{2}$; however, via use of manipulatives we find that the best unit for representing both fractions in the problem is sixths. If we are using $\frac{4}{6}$ and $\frac{3}{6}$ and a single set of manipulatives, we may find that we run out of manipulatives, as there are not enough sixths in one set to represent both fractions. There are a number of ways to resolve this issue: we could combine fraction sets with a partner, so that we see the sum is $\frac{7}{6}$. Or we could re-represent the first group of 6 sixths using a one-whole piece, and then add one more, so the answer is $1\frac{1}{6}$. Yet another idea is to draw the sixths on paper. It is interesting for students to share their methods for problems like this, and to compare the formats of the answers (e.g., $1\frac{1}{6}$ and $\frac{7}{6}$). (Note that, in this case, twelfths could have been used, providing yet another variation on the method of the solution and format of the answer.) Again, many contexts that involve combining amounts are possible.

4. Both fractions in this problem have the same denominator, so it is not necessary to re-represent them. The operation here is subtraction—hence, typically the $\frac{3}{5}$ is modelled, and then the $\frac{1}{5}$ is *removed from* it. Alternately, if we model both amounts, subtraction can be viewed as the *difference*, or the amount added to $\frac{1}{5}$ to get $\frac{3}{5}$. In either case, the answer is $\frac{2}{5}$. While we could also choose to re-represent the fractions using another unit, such as tenths, this approach does not add anything to the solution quality, although it is not wrong, nor should it be indicated to students that it is wrong. Any contexts that start with $\frac{3}{5}$ and use the idea of removing (eating, giving away, spending time, separating out, etc.) $\frac{1}{5}$ are appropriate.

5. To solve this problem, it is necessary to explore fractional units until one is found that fits both fractions, which turns out to be tenths. Once the $\frac{3}{5}$ is re-represented as $\frac{6}{10}$, the problem can be completed using the methods described for Problem 4, yielding an answer of $\frac{3}{10}$.

6. A number of possible routes can be taken to solve this problem. If plastic manipulatives are used, then we will need more than the 4 fourths in one package. As described above, we could combine two packages, re-represent the $\frac{5}{4}$ as one whole and $\frac{1}{4}$ ($1\frac{1}{4}$), or use a drawing instead. We then need to represent the fourths as eighths (either $1\frac{2}{8}$ or $\frac{10}{8}$) and then proceed as in Problems 4 and 5, which will yield a final answer of $\frac{7}{8}$. However, for this problem, there is yet another strategy available to us. We could think of the $\frac{5}{4}$ as $1\frac{1}{4}$, and set aside the $\frac{1}{4}$ to begin. The remaining part is $1 - \frac{3}{8}$, which is more easily seen as $\frac{5}{8}$. The problem now becomes $\frac{1}{4} + \frac{5}{8}$. To complete the problem, $\frac{1}{4}$ needs to be represented as eighths, which yields the same final answer of $\frac{7}{8}$. Contexts similar to those used in Problems 4 and 5 will also work for this example, but the initial amount needs to be greater than one whole, for example, describing a situation in which there are 5 slices of pizza and each pizza is cut into fourths.

7. Any of the methods and contexts described in Problem 6 will also work for this problem. We could directly re-represent each value using a unit that works for both fractions—in this case, tenths—and then subtract, yielding $\frac{9}{10}$ as the answer. Alternately, we could think of $\frac{7}{5}$ as $1\frac{2}{5}$, set the $\frac{2}{5}$ aside for now, and subtract the $\frac{1}{2}$ from the 1. The problem now becomes $\frac{1}{2} + \frac{2}{5}$, and again, tenths are needed to re-represent both fractions and add. We see that by adding $\frac{5}{10} + \frac{4}{10}$ we reach the same answer as we did using the other methods.

Extensive hands-on work allows students to develop *fraction-sense*, which includes a sense of the size of various fractions, and a notion of what units work to re-represent fractions. Such informal work also makes it clear that *reasoning* is the most important aspect of problem solving—it is "the math." Importantly, modelling makes problems involving fractions conceptually accessible to students, rather than mysterious. And lastly, it is these concrete models, together with reasoning, that will be used (as we will see in the sections to come) to develop the more formal numeric or symbolic methods.

6.6 Construction of the Common Denominator

When students are working concretely to re-represent fractions (for example, moving from halves and thirds to sixths), they might be asked, "How did you know to use sixths?" While at first children will simply use trial and error, after a while they will begin to recognize a connection between the denominator values. Only when such a connection begins to emerge should students begin to explore more formal methods.

As mentioned above, to find a unit that works to add or subtract fractions with unlike denominators, we can *always* multiply the denominators. For example, when adding $\frac{1}{2}$ and $\frac{1}{3}$ we can use sixths, and for $\frac{2}{3}$ and $\frac{3}{4}$, we can use twelfths. We observe that $2 \times 3 = 6$ and $3 \times 4 = 12$. While this method always yields a denominator that works, it does not always generate the simplest unit. In more formal language, this method yields a *common denominator*, but not necessarily the *lowest common denominator*. For example, when adding $\frac{1}{2}$ and $\frac{1}{4}$, the *multiply the denominators* method suggests using eighths. While eighths *can* be used to add $\frac{1}{2}$ and $\frac{1}{4}$ they are not the smallest possible, or lowest possible, common denominator.

Exploration/Task

Draw a model on paper showing how you might solve $\frac{2}{3} + \frac{1}{4}$. Be sure to draw both of the initial fractions, showing each as a shaded part of a separate one whole, as well as how you might re-represent each in order to subsequently perform the addition. Think about how you are using your model to construct the new unit.

Follow-Up and Discussion

When completing the task above, you likely found yourself re-representing the $\frac{2}{3}$ as $\frac{8}{12}$. If you were using paper fractions or drawings, you may have noticed that you were cutting the thirds into 4 *times* as many pieces. Similarly, you might have noticed that you cut the fourths into 3 *times* as many pieces, to get $\frac{3}{12}$. In each case, you generated twelfths. In each case, because the size of the piece is now smaller, you also need *more* of them.

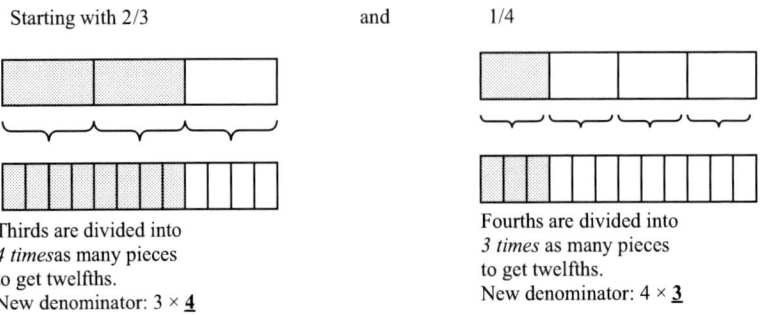

Trying a similar process of subdividing with other unlike denominators allows the generalization to emerge that multiplying the denominators together to find the new denominator will always work. Note that both the 3 and the 4 clearly divide into 12, which is the *product* of 3 and 4. Using formal language, both 3 and 4 are *factors* of 12, and 12 is a *multiple* of each.

> **CONNECTION**
> For descriptions of the concepts of factors and multiples, see section 5.6.

Exploration/Task

Test the idea of multiplying denominator values to find a common denominator for $\frac{1}{2} + \frac{2}{5}$, then think about how you could find the new numerator of each. Find the sum numerically, and check your thinking using manipulatives.

Follow-Up and Discussion

If we try finding a new denominator for $\frac{1}{2}$ and $\frac{2}{5}$, we see that a new denominator can be generated from 2×5, which is 10. It works to re-represent both $\frac{1}{2}$ and $\frac{2}{5}$ as tenths, because numbers exist that multiply each of 2 and 5 to get 10. In other words, 10 is a *multiple* of both 2 and 5, and we *multiplied* 2 and 5 to generate the 10.

This method is an important and useful one. Multiplying will *always* generate a unit that works (a *common denominator*). If we have three or more different fractions with unlike denominators, we can still multiply all of the denominator values to generate a new denominator unit. For example, to add $\frac{1}{2} + \frac{1}{3} + \frac{3}{5}$, since $2 \times 3 \times 5 = 30$, we know we can use thirtieths (or 30) as a common denominator. Looking ahead, such knowledge is important in algebra when techniques must be developed to solve expressions with denominators.

We will now explore a further refinement of this initial numeric method of multiplying denominators. This refinement not only always works, but also allows us to find the *best* (simplest) possible value for the denominator, which the previous method of simply multiplying does not always do.

From our previous hands-on work, we have observed that, for example, $\frac{1}{6} + \frac{1}{4} = \frac{2}{12} + \frac{3}{12} = \frac{5}{12}$. Yet using the numeric method described above, we might have chosen to use 24 as a common denominator instead of 12, since $6 \times 4 = 24$. How can we know whether a simpler common denominator (than the product of the denominators) exists, and if it does, how can we determine what it is?

Exploring this question might make an excellent problem-based classroom lesson. The critical conceptual element to the notion of a common denominator, or unit that works to re-represent all fractions, is that the new denominator must be a *multiple* of each denominator value. But we would like to find the *lowest* such multiple. The lowest multiple is the lowest (or smallest possible) common denominator.

For example, if we want to add halves and sixths, we can cut the halves into *3 times as many* pieces, to re-represent them as sixths:

> **INSIGHT**
> Fractions need to be represented in the same unit size before we can add or subtract them symbolically.

> **INSIGHT**
> Multiplying a denominator by a value is like cutting one whole into that many times smaller pieces.

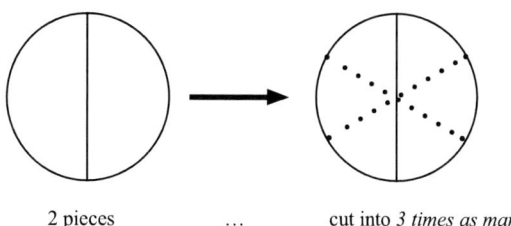

2 pieces ... cut into *3 times as many*

A *common denominator* is *any* new denominator that is a multiple of all fraction denominator values initially being added or subtracted. The *lowest common denomi-*

nator is the smallest of these possible values. To find this lowest possible denominator, we are looking for the smallest number that is a *multiple* of *all* of the original denominator values.

Exploration/Task

Use manipulatives and reasoning to explore the simplest denominator that can be used to add $\frac{1}{2} + \frac{1}{6} + \frac{3}{4}$. Then use models to represent both the initial fractions and their equivalent common denominator forms.

Follow-Up and Discussion

> **CONNECTION**
> To review the concept of multiples—numbers that are products of a given number—see section 5.6.

Using manipulatives, twelfths work to add these fractions. If we had used the numeric method of multiplying the denominators together we would have found a common denominator of 48 (the product of 2 × 6 × 4). But the *lowest* common denominator is 12, because it is the smallest multiple. Other possible denominators are 24 and 36, because each of these numbers is a multiple of all three denominators—2, 6, and 4.

We now know that sometimes a new denominator exists that works as a common denominator that is smaller than the one found by multiplying each of the denominators of the fractions in the problem. But how can we find this value efficiently?

Recall that a number that is a multiple of two or more other numbers is called a *common multiple*. The smallest of those numbers is called the *lowest common multiple*. For example, if we compare multiples of 4 (4, 8, 12, 16, 20, 24…) and 6 (6, 12, 18, 24…), we find that the first two common multiples are 12 and 24. In the case of 4 and 6, the lowest common multiple is 12, as it is the first or smallest multiple of these two numbers.

> **INSIGHT**
> The lowest common denominator is simply the lowest common multiple.

If students have studied the concept of the *lowest common multiple* when learning about whole numbers, then, in fact, they already have a procedural method of finding the *lowest common denominator*. If not, it is time to address the concept. One effective visual way of finding the lowest common multiple is to examine a list of multiples of each denominator, and look for *common* ones. (Sometimes students struggle with the meaning of the word *common*; initially, it might be clearer to say, "multiples that are the same.")

Example

To find the lowest common multiple of 6 and 15 we write:

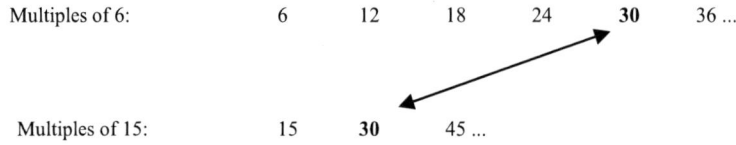

Follow-Up and Discussion

The first multiple that 6 and 15 share is 30—it is their lowest *common* multiple. This number is the smallest possible denominator we could use to add or subtract fractions involving sixths and fifteenths, hence it is the *lowest common denominator*.

If we kept going, we would also find 60 and 90 are common multiples, hence these other (larger) common denominators could also be used.

> **MATHEMATICAL TERM**
>
> The *lowest common multiple* of two or more values is the first or smallest value that they share as a multiple, or, in other words, that is a multiple of each.

Example

A more visual illustration of the first shared multiple or lowest common multiple concept for $\frac{1}{4} + \frac{1}{6}$ is illustrated below:

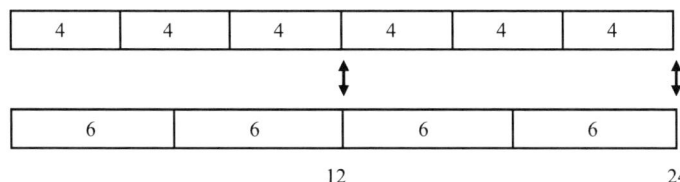

Follow-Up and Discussion

We might use 4 and 6 centimetre strips or rods, cubes, or lengths measured with rulers to build this model. This model (sometimes called the *trains* model, since we are building "trains" of the length of each value) illustrates the numeric lengths of multiples of 4 and 6. The trains show that 12 and 24 are multiples of *both* 4 and 6—can you see why? Twelve is a common multiple of 4 and 6, and is in fact the lowest or smallest one. Note that if we had simply multiplied 4 × 6, we would have gotten 24. Although multiples of 4 and 6 line up at both 12 and 24, 12 is the *first* place that they align, so 12 is the *lowest* common multiple. Thus it is also the lowest common denominator.

Being able to numerically predict the best possible denominator for, say, fourths and sixths allows us to reach for the manipulatives from the twelfths set with confidence, and without using trial and error; however, we would still need to use the manipulatives to figure out *how many* twelfths we require to add the fourths and sixths. In the next section, we will explore a numeric method to determine the correct numerator of the new (equivalent) fractions.

Practice and Further Exploration

Use the trains model to find the smallest common denominator for 6 and 9.

6.7 Developing Procedures for Adding and Subtracting Fractions

The previous section established the idea that fractions can be readily added or subtracted once the denominators or units are the same, as well as numeric ways to find a new denominator. We now need to examine methods of numerically finding the appropriate *numerators* for the equivalent fractions. After completing this phase, we will have developed a numeric method for adding and subtracting fractions—with understanding!

In an example used in section 6.6, we saw how thirds and fourths might be re-represented as twelfths to allow us to add or subtract. So far we have been able to identify the *denominator* using the lowest common multiple idea, but we still needed manipulatives to find the *numerator*. For example, for $\frac{2}{3} + \frac{1}{4}$, we know to use twelfths, but not *how many* twelfths.

Exploration/Task

Examine the process of changing $\frac{2}{3}$ to twelfths. Pretend you are using paper strips and scissors as shown below. Refer to the diagrams and answer the guiding questions to think through the process.

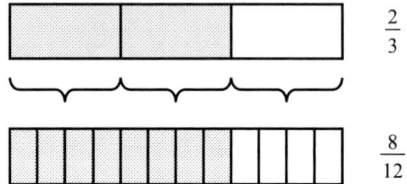

Guiding questions:

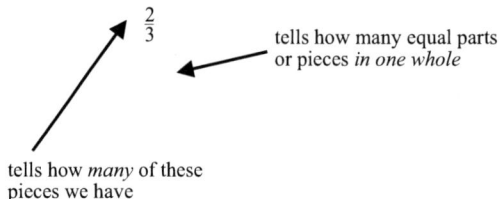

1. Recall that the 3 (the denominator) tells us how many pieces the whole is cut into. How many *times the number of pieces* do we need for the whole to be cut into twelfths?
2. The 2 in the fraction numerator originally told us that we had 2 pieces (of size one-third). How many *times the number of pieces* do we need to shade or select, if the pieces are now size one-twelfth?

As you work on this task, remember that we are trying to find a *numeric* method to change $\frac{2}{3}$ to $\frac{8}{12}$. We would like to generate such a method so that we can use it to solve other problems in the future, instead of using drawings or manipulatives.

Follow-Up and Discussion

You might have noticed that as we move from thirds to twelfths, the pieces are cut or divided into *4 times as many* parts.

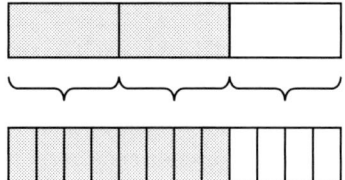

Both the number of pieces into which the one whole is cut (the denominator), and the number of pieces we actually have (the numerator), are 4 times as great. Similarly, there are 3 times as many pieces in the numerator and denominator of $\frac{1}{4}$, when changing fourths to twelfths. Each one-fourth piece is re-represented in 3 times as many pieces. These pieces are smaller—they are one-third the original size.

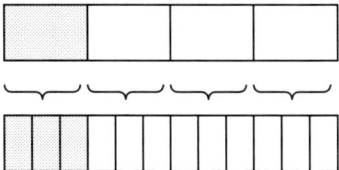

> **INSIGHT**
>
> When we take a fraction such as $\frac{2}{3}$ and cut the 2 one-third pieces into 4 times as many pieces, it doesn't change the total amount of what we have, as long as we remember to cut the denominator unit size into 4 times as many pieces too, and take more of them. The size of the pieces will be smaller than before, but we will have more of them.

Note that in each case this process does not change the overall value of the fraction; we are just making the pieces *smaller*, and thus have to choose *more* of them, by multiplying by the same value.

This thinking can be recorded more formally using the fraction notation. It is important to remember the roles of the two numbers in any fraction—the bottom number tells us the number of pieces in one whole, and the top number tells how many pieces we have—as we do this.

$$\frac{2}{3} = \frac{2 \times 4}{3 \times 4} = \frac{8}{12}$$

We emphasize that the above reasoning relies *only* on the already established meaning of the numerator and denominator in the part-whole fraction notation (which we originally simply called the top and bottom numbers of the fraction) (see section 6.1). The only multiplication concept required here is whole number multiplication, i.e., 2×4 and 3×4 (along with the fraction definition).

Another explanation for the equivalent fraction procedure that is sometimes imposed on students involves telling them to "multiply $\frac{2}{3}$ by $\frac{4}{4}$ because that is 1, and multiplying by 1 doesn't change the number." Although students may later be able to see that multiplying by $\frac{4}{4}$ has the same numeric *result* as $(2 \times 4)/(3 \times 4)$, we advise *against* using or suggesting the method $\frac{2}{3} \times \frac{4}{4}$ at this developmental stage for two important reasons. First, it is difficult for students to understand the conceptual basis for why we would want to multiply by something that is "equal to 1 so it won't change the number." This method is one of those tricks that most students do not understand. Worse still, it is difficult to conceptually justify, so when they question why this approach is being used, there may not be a good answer. Second, there is a problem in the developmental progression here. Most children at this stage do not know the numeric method for multiplying two fractions, because in most curricula multiplication of fractions is *not* introduced prior to addition. Imposing without justification that $\frac{2}{3} = \frac{2}{3} \times \frac{4}{4} = \frac{8}{12}$ requires students to accept a rule about multiplying that they likely have no conceptual or procedural basis for understanding, and must therefore just memorize. Sense-making is lost.

We suggest using the method explored earlier in this section, which relies on the definition of the fraction, in order to see that we are changing the value of the 2 and 3 in $\frac{2}{3}$ by simply cutting the $\frac{2}{3}$ into *4 times as many* pieces (using multiplication by 4), which we then see is (re)named $\frac{8}{12}$.

Exploration/Task

Using the method described in the previous task, consider the process of re-representing $\frac{3}{4}$ as eighths. Establish the numeric steps using paper strips (or models of them) and reasoning similar to that used above to represent $\frac{2}{3}$ as $\frac{8}{12}$. Here is a model to get you started:

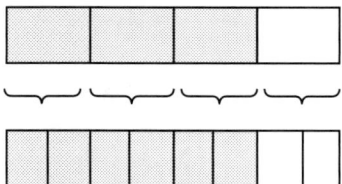

Follow-Up and Discussion

In the case of fourths and eighths, you should have found that you were cutting or subdividing the fourths into *twice as many* pieces as before. Numerically, this is like multiplying both the numerator and denominator values of $\frac{3}{4}$ by 2.

Example

Suppose we want to add $\frac{2}{3}$ and $\frac{1}{4}$ using a numeric method. Perhaps we have one can of paint that is $\frac{2}{3}$ full, and another can of the same size that is $\frac{1}{4}$ full, and we want to know how much paint we have in total.

Using the methods outlined in the two previous tasks in this section, we can find two fractions with like denominators that can easily be added. This allows us to complete the entire sum numerically. The procedure is illustrated below:

$$\frac{2}{3} + \frac{1}{4} = \underbrace{\frac{2 \times 4}{3 \times 4}}_{\text{cut 4 times as many times}} + \underbrace{\frac{1 \times 3}{4 \times 3}}_{\text{cut 3 times as many times}} = \frac{8}{12} + \frac{3}{12} = \frac{11}{12}$$

After identifying 12 as the lowest common multiple of 3 and 4, we need to re-represent each fraction as twelfths. Note that this numeric method aligns exactly with the idea of the fractions being cut into more pieces, as explored earlier. Previously, we cut the thirds into 4 times as many pieces. Numerically now, to change thirds to twelfths, we might think *3 times what is 12?*

Reasoning

By "cutting" fractions into smaller pieces—by multiplying the denominator by a value—we have established a numeric method or symbolic procedure for rewriting fractions with different denominators so that they have the same denominator. Establishing common denominators allows us to numerically add or subtract the fractions. The conceptual basis for this method involves the meaning of the fraction notation itself, as well as the idea of cutting fractions into more pieces. In the following task, you will have the opportunity to try out these ideas.

Exploration/Task

Using the reasoning presented in this section, draw models of each operation and show how you would re-represent each fraction to carry out the numeric procedure. In each case, draw a concrete model, and work through the steps using your model and reasoning as you did to solve previous problems. Next, write out each question using the numeric method. Think about how the model helps you understand the meaning behind the numeric method each time you use it. Finish each question by calculating the final answer.

1. $\frac{2}{5} + \frac{3}{10}$
2. $\frac{2}{3} - \frac{1}{4}$
3. $\frac{5}{6} - \frac{3}{4}$
4. If you had chosen twelfths as the new denominator for the third question, you likely got $\frac{1}{12}$ as an answer. But what if a student had not recognized twelfths as the *lowest* common denominator, used the "multiply the denominators together" method, and hence used twenty-fourths? The answer would likely have been $\frac{2}{24}$. Replicate the student's reasoning and solution.

Follow-Up and Discussion

Modelling your solution first allows you to compare your answer to the one that you found using the numeric method—of course, your two final answers should agree! Students sometimes find different equivalent fractions as answers. For example, two students with different but equivalent answers, such as $\frac{1}{12}$ and $\frac{2}{24}$, as might be found for questions 3 and 4 respectively, might compare their thinking (as illustrated below), to determine if their answers are the same.

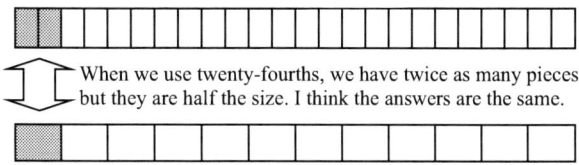

When we use twenty-fourths, we have twice as many pieces but they are half the size. I think the answers are the same.

The situation in which students are using models, and not yet simplifying final answers, presents a dilemma for those of us trained to always convert fractions to *lowest terms* using a formal numeric method. For fractions that can be relatively easily represented with models, we recommend that informal reasoning be used to determine if fractions are equivalent, and not to push for a particular numeric form to soon. This allows students to begin to make observations about simpler ways to write amounts on their own, using reasoning. The final section of the next chapter (section 7.8) further examines numeric methods for converting fractions to other equivalent forms, including the form called *lowest terms*. Initially, students should be encouraged to leave their answers in whatever form makes most sense to them.

CONNECTION

For the method of converting fractions to lowest terms, see section 7.8.

Chapter Problems

1. Temporarily suspend any knowledge you have of numeric methods for fraction calculations. Solve the following using only concrete manipulatives or a drawn model, while thinking of the reasoning a student might use. If possible, use different reasoning to solve each problem in more than one way.

    a) $\frac{3}{4} + \frac{5}{4}$
    b) $\frac{1}{2} + \frac{3}{8}$
    c) $\frac{5}{6} - \frac{3}{6}$
    d) $\frac{5}{6} - \frac{1}{3}$

2. Re-solve the above questions using a numeric method as developed in this section. In each case, compare your solution to your model.

3. Presenting contextual problems to students is important to developing understanding. Solve the following problem as a student might, using only models and reasoning: Ben's novel has 100 pages. He has read $\frac{3}{4}$ of the novel. Daniel's novel has 120 pages. He has read $\frac{2}{3}$ of the novel.

    a) Who has read more pages?
    b) Who has more pages left to read?

Further Reading

Phelps, K. A. G. (2012). The power of problem choice. *Teaching Children Mathematics, 19*(3), 152–157.

> Phelps uses differentiated instruction strategies to help students understand the concept of adding and subtracting fractions that do not have a common denominator.

Polly, D., ed. (2012/2013). The unusual baker. *Teaching Children Mathematics, 19*(5), 286–289.

> Polly discusses the experiences of elementary students learning basic concepts of fractions, sharing their solutions and class discussions.

Wilkerson, T. L., Bryan, T., and Curry, J. (2012). An appetite for fractions. *Teaching Children Mathematics, 19*(2), 90–99.

> Wilkerson, Bryan, and Curry detail how to use chocolate bars to help Grade 6 students gain an understanding of fractional concepts with denominators that are a factor of 12.

Chapter 7
Multiplicative Fraction Representations and Operations

7.1 Fraction Multiplication: Concept and Contexts

As we saw with whole number addition and multiplication, while there is a relationship between additive and multiplicative reasoning, they also have distinctive features.

If students understand multiplication only as repeated addition, or if they have been told that multiplication results in an answer that is larger than the two numbers being multiplied, it is important to broaden their general understanding of multiplication before exploring fraction multiplication. These very narrow interpretations, which teachers may think are helpful with respect to whole numbers, can make fraction multiplication seem confusing and counterintuitive. As with other operations, it is important to have a deeper and more flexible understanding of multiplication before moving on to learn about fraction multiplication. Context continues to play an important role in helping students learn the concepts. The sample problems below will help establish models for multiplying fractions.

> **CONNECTION**
> To review whole number multiplication models and procedures, see sections 4.6 and 5.2.

Exploration/Task

Answer these questions using only models and reasoning. The purpose is to think about the interpretation of the operation as it is used in each case.

1. Cupcakes come in packages of 6. If 3 packages are purchased, how many cupcakes have been bought? Draw a model of this problem and use it to illustrate your reasoning. What might be a good name for the strategy or operation you used?
2. One of the packages of 6 cupcakes was sitting on the counter when Donald came home. His mother told him he could eat half the cupcakes in the package. How many cupcakes could he eat? Draw a model to illustrate your reasoning, and solve the problem using this model. Compare the reasoning

used here to the reasoning you used in the previous question. How does the operation compare? Is there more than one operation that could be used to solve this problem?

Follow-Up and Discussion

As you solved the first problem, you may have found yourself using a repeated addition strategy. The answer (3 packages of 6, or 18 in total) is a value greater than a complete package of 6. The idea of *3 of 6* might have been interpreted as 3 × 6; it could also have been expressed as 6 + 6 + 6.

To solve the second problem, you might also have started by drawing a package of 6, and then operated on that package of 6 to show *half of* it. It is difficult to apply a repeated addition strategy to find half of something—we really do need to think multiplicatively, not additively, to find *half times as much*. Your second model may have looked like this:

> **INSIGHT**
> Multiplying doesn't always result in an answer that is greater than the initial amount.

If students who are presented with a problem like this are unable to identify the *half of* interpretation as $\frac{1}{2}$ *times*, encourage them to think about the connection between the *3 of* and *half of* in each of the above problems.

Exploration/Task

Answer the following problem using only models and reasoning:

Jason usually runs 8 laps of the track after school for exercise. Today he was feeling great and he was able to run one and a half times his usual distance. How many laps did he run?

Did Jason run more or less than usual today? Think about the interpretation of multiplication you used to come to your conclusion.

Follow-Up and Discussion

Sometimes comparing *one and a half times* with *two times* (or *two of*) helps students see both the model and the multiplicative operation required. It may also help to think of *one and a half of* as *one* of Jason's usual routine (8 laps) plus another *half* of it (4 more), which tells us he ran 12 laps in total today.

It is important to continue to offer students problems in context until they are able to identify the multiplication concept in the problems. They need to be presented with problems in which the result is both greater and less than the initial value provided. If students are merely *told* that the problems are "multiplication problems," they will not develop the understanding needed to identify multiplication as the operative concept when presented with a new problem for which they do not know the operation. This common mistake, which is usually made in an attempt to make things easier for students, is an example of why word problems typically cause students angst. By suggesting or imposing a method and operation on them, we rob them of the chance to actually *think* mathematically.

Exploration/Task

Use a model and reasoning to answer the following problems. Think about what operation you are using, and how you decided to do so. After answering each question using your model and reasoning, try to represent the situation using a numeric statement of a mathematical operation. Solve each problem using the numerical operation.

1. To cover the floor of the dance studio for stretching class, 8 large mats are needed. The mats completely cover the floor, and are all the same size. If the floors of 2 studios of the same size are both covered with the same type of mats, how many mats are needed in total?
2. For stretching class at the dance studio, 6 large mats are placed on the floor. The mats cover three-quarters of the floor, and are the same size. One-third of the mats on the floor are black. How much of the floor was covered by black mats?

Follow-Up and Discussion

In the first problem, the concept of 2 studios might be interpreted as whole number multiplication by 2, resulting in 16 mats in total. In this case, repeated addition, as well as other models such as an area model, could be used to interpret the problem. In the second problem however, you might have drawn something similar to this:

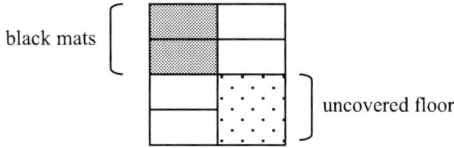

We can begin by identifying the area covered with mats, which is three-quarters of the total area. By using the model and mentally grouping the area covered with mats into 3 equal sections, we find that *one-third of* the mat area (2 of the 6 total mats) is covered with black mats. (The other 4 mats represented in the model are white.) We can conclude by looking only at the black mats in the model that the black mats cover one-quarter of the *total* floor. The complexity of this problem is due to the fact that we need to keep two quantities in mind, both the one whole of the floor, and the three-quarters of the floor that is covered in mats. The black mats are simultaneously one-third of the mat area, and one-quarter of the total floor area.

The model indicates that one-third of the three-quarters is one-quarter. After writing out this phrase, you might also have been able to connect your model to this numeric calculation: $\frac{1}{3} \times \frac{3}{4} = \frac{1}{4}$. Recall that three-quarters of the floor is covered by mats, and one-third of that area is covered by black mats. Reviewing the model makes it clear that the black mats cover one-quarter of the total floor.

Practice and Further Exploration

Use models and reasoning to represent and solve each of the following problems:

1. It took Sam three-quarters of an hour to mow 1 lawn. He mowed 3 similar lawns. How long did it take him to mow all 3?

2. On Saturday, Sam was given the job of mowing the lawn at the resort where he works. The job usually takes one and a half hours. After he had completed two-thirds of the job, he stopped for a drink. How long had he been mowing for at that point?
3. When Sam got home there were 8 slices of a large 12-slice pizza left on the counter. Sam was so hungry that he ate half of what was there. What fraction of the original pizza did he eat?

7.2 Concrete Models and Strategies for Multiplying Fractions

In this section, we further explore multiplication with fractions by using models and connecting to number statements.

Multiplying a Fraction and a Whole Number

Recall that one way to think of multiplication is as working with groups of an amount. So "5 times 6" can mean 5 groups *of* 6 or, simply, 5 *of* 6. Multiplying by a fraction that is less than one results in only part of a group.

Exploration/Task

Create as many models as possible to illustrate multiplying a fraction by a whole number. Use the example $\frac{1}{2} \times 6$, which might arise in the context of groups (mentioned above) or when thinking about part of a package of 6 cookies.

Follow-Up and Discussion

We can read the statement $\frac{1}{2} \times 6$ aloud as *one half of 6*. Below is one possible line of reasoning, shown using chips or counters that could be used to model the solution.

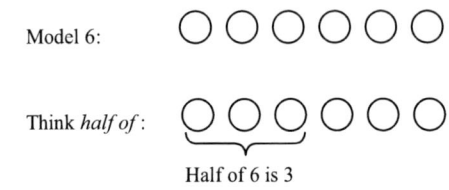

Another potential interpretation of *one half of 6* involves the number line. We could represent 6 on a number line as a distance, and could think of *half the distance*.

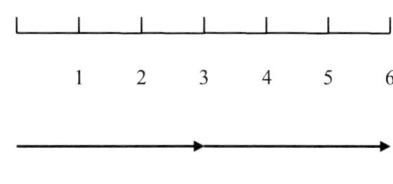

Half the distance to 6 is 3

Reasoning

Taking a fractional amount of a whole number is like multiplying, but we don't always get an answer that is greater than the number we started with.

Multiplying Two Fractions

What happens when we multiply two fractions? How can we model and make sense of this type of operation?

Example

Model $\frac{1}{3} \times \frac{3}{4}$. This calculation might emerge from a context such as, "There was three-quarters of a pizza left. If you ate one-third of the remaining pizza, what fraction of the original whole pizza did you eat?" Below is one possible model that could be used to solve this problem.

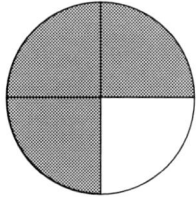

One way to interpret $\frac{1}{3} \times \frac{3}{4}$ is to think *one-third of three-quarters*, so the model begins with three-quarters.

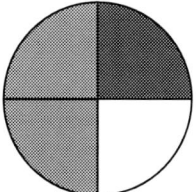

One-third of the 3 pieces (each of which is size one-quarter), is 1 piece (of size one-quarter). Therefore, one-third of three-quarters is one-quarter. Numerically, $\frac{1}{3} \times \frac{3}{4} = \frac{1}{4}$.

For this example, we were able to directly model the operation without representing the three-quarters (the fraction to be multiplied) as another fraction using a unit smaller than quarters. But what if the number of pieces in the fractions doesn't allow us to easily model the multiplication operation? The following example illustrates this situation.

Example

Model $\frac{3}{4} \times \frac{2}{5}$. Can you think of a suitable context that might give rise to this question? This time, we will use fraction strips to illustrate how a different manipulative can be used. As we did previously, we can think *three-quarters of two-fifths*. First, model $\frac{2}{5}$:

There are only 2 pieces (2 of size one-fifth), so we can't immediately take three-quarters of the 2 pieces. We would need to have 4 pieces to think *three of the four*, or *three-quarters of* the amount. One idea is to represent the $\frac{2}{5}$ as tenths:

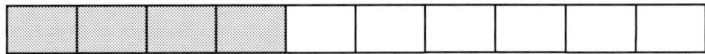

Using the models, we can see that $\frac{2}{5}$ is equivalent to $\frac{4}{10}$.
Using the new representation, we can think of "$\frac{3}{4} \times$" as "three of the four pieces." We can now use the model to complete the multiplication operation.

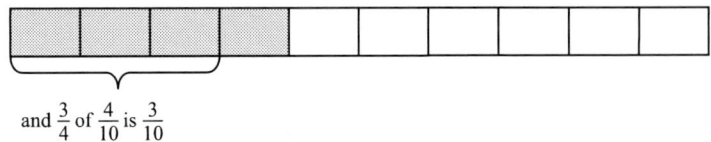

and $\frac{3}{4}$ of $\frac{4}{10}$ is $\frac{3}{10}$

We find that $\frac{3}{4} \times \frac{2}{5} = \frac{3}{10}$. Other types of models, such as area models, could also be used to solve this type of problem. We will look at this in greater detail later in this section.

> **CONNECTION**
> To review how to represent a given fraction as an equivalent fraction, see section 6.4.

Fractions Greater Than One

What if you are multiplying two fractions and one of them is greater than 1? Let's explore.

Example

Model $1\frac{1}{2} \times \frac{2}{3}$. The following is an example of an appropriate context: "A recipe for 2 dozen cookies calls for $\frac{2}{3}$ cup of sugar. To make 3 dozen cookies, you need to use $1\frac{1}{2}$ batches of the recipe. How much sugar do you need?" One interpretation of this problem is to think *one and a half portions* of two-thirds. Start with a model of $\frac{2}{3}$:

One and a half of this amount (two-thirds) is like *one* complete group or portion of $\frac{2}{3}$, and *half* of another portion of $\frac{2}{3}$:

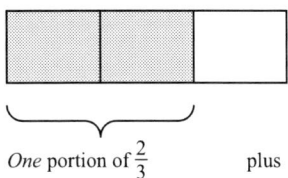 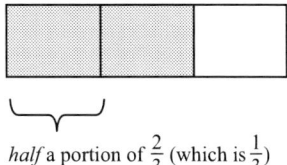

One portion of $\frac{2}{3}$ plus   *half* a portion of $\frac{2}{3}$ (which is $\frac{1}{3}$)

$\frac{2}{3} + \frac{1}{3}$ gives $\frac{3}{3}$ all together ... which happens to be 1.

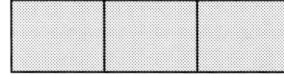

Numerically, $1\frac{1}{2} \times \frac{2}{3} = \frac{3}{3} = 1$.

> **INSIGHT**
>
> The product of fraction multiplication may result in *more* than we started with (if we are multiplying by an amount more than 1) or *less* than we started with (if we are multiplying by less than 1).

Fraction Products and Area Models

If the fractions need to be re-represented in more pieces to model the operation, and the way in which they should be re-represented isn't obvious, an area model is helpful.

> **CONNNECTION**
>
> To review the use of area models in whole number multiplication, see section 5.2.

Example

Model $\frac{2}{3} \times \frac{4}{5}$. To model and solve this problem, we will use an area model. Area models can always be used, but are not necessary for problems such as $\frac{1}{3} \times \frac{3}{4}$, which can be solved without re-representing the $\frac{3}{4}$. For this example, since $\frac{2}{3} \times \frac{4}{5}$ cannot be solved without re-representing the $\frac{4}{5}$, an area model is helpful. The issue here is that $\frac{4}{5}$ cannot be directly split into 3 pieces, as would be needed for the "two-thirds of" operation.

We start with a model of $\frac{4}{5}$. Think *two-thirds of the four-fifths*. One way to interpret the situation is to take two-thirds of *each* one-fifth. To do this, we have to re-repre-

> **CONNECTION**
>
> To review the construction of equivalent fractions, see sections 6.3 and 6.4.

sent each fifth as 3 (smaller) pieces; that is, we must break each fifth into "thirds" of the fifth. We are effectively re-representing the whole as *three times* as many smaller pieces.

To multiply $\frac{4}{5}$ by $\frac{2}{3}$, we can think about taking two-thirds of *each* one-fifth. In our model, this is like shading 2 of each column of 3 boxes. The patterned area—8 pieces of the 15 total pieces in this whole—represents the answer, which is $\frac{8}{15}$ of the whole. Numerically, $\frac{2}{3} \times \frac{4}{5} = \frac{8}{15}$.

Practice and Further Exploration

1. Create a classroom fractions task with a context that requires a model of fraction multiplication but does not involve re-representing the fractions. Model and solve your question.
2. Create a fractions task with a context that requires a model of fraction multiplication that does involve re-representing one of the fractions. Model and solve your question.

7.3 Developing Procedures for Multiplying Fractions

Practice with hands-on materials and drawn models provides students with critically important experience that helps them make sense of fraction multiplication procedures. Students who have gained enough experience with models *know* that two-thirds of, say, three-fifths, is two-fifths because they can visualize it; they *see* three-fifths in their mind, and then use visual reasoning to quickly imagine two-thirds of those 3 pieces—which has to be two-fifths. Such experienced students have no need to use numeric procedures for such simple calculations.

However, we also saw in the previous section that not all fraction multiplication questions can be performed without re-representation. For example, we cannot take $\frac{3}{4}$ of $\frac{2}{5}$ directly. We can, however, easily substitute $\frac{4}{10}$ for $\frac{2}{5}$, and $\frac{3}{4}$ of $\frac{4}{10}$ can be calculated visually—the answer is 3 of the $\frac{4}{10}$, or $\frac{3}{10}$. Numerically, we can write: $\frac{3}{4} \times \frac{2}{5} = \frac{3}{4} \times \frac{4}{10} = \frac{3}{10}$.

Exploration/Task

Solve the following examples using (mental) visual reasoning; do not use any formal procedure. You may need to identify a simple equivalent fraction, as in the above example.

1. $\frac{1}{2} \times \frac{2}{3}$
2. $\frac{1}{3} \times \frac{6}{8}$
3. $\frac{2}{3} \times 1\frac{1}{2}$
4. $\frac{3}{4} \times \frac{2}{3}$

Follow-Up and Discussion

The easiest way to determine if a problem can be solved using visualization is to simply try it. For the first problem, if you imagine 2 pieces of size $\frac{1}{3}$ floating around in your mind, half of this amount is 1 of these thirds. Similarly, for the second problem, we can visualize $\frac{1}{3} \times \frac{6}{8}$ by mentally arranging the 6 pieces (eighths) into 3 equal groups, so that there are $\frac{2}{8}$ in each group. Finding $\frac{1}{3}$ of the amount is done by taking 1 of these 3 groups (of $\frac{2}{8}$): the answer is $\frac{2}{8}$. In the third problem, $1\frac{1}{2}$ can be visualized as 3 pieces of size one-half. Two-thirds of the amount is 2 of the pieces of size $\frac{1}{2}$, or 1. For the final problem, however, we cannot take $\frac{3}{4}$ of the $\frac{2}{3}$ without using an equivalent fraction for the $\frac{2}{3}$, for example, $\frac{4}{6}$. Taking $\frac{3}{4}$ of the $\frac{4}{6}$ gives an answer of $\frac{3}{6}$, which can also be thought of as $\frac{1}{2}$.

When a teacher has become more experienced, it becomes possible to know which questions will be easier than others for students, i.e., whether equivalent fractions are needed or not. The key is the relationship between the denominator of the first fraction, and the numerator of the second. They tell us whether we can form the groups necessary to solve the problem. For example, examine $\frac{2}{3}$ of $\frac{6}{8}$:

$$\frac{2}{3} \quad \times \quad \frac{6}{8}$$

Here, we can arrange the 6 pieces (of size $\frac{1}{8}$) into 3 equal groups. We can take two-thirds of the amount without subdividing the $\frac{6}{8}$. Note that 6 is divisible by 3.

At some point, drawing models for every single fractions product becomes too tedious. It would be helpful to have a numeric method that can be used instead. We will develop such a method using an example.

Example

At the end of section 7.2, an area model was used to illustrate $\frac{2}{3} \times \frac{4}{5}$. Subdivision of the $\frac{4}{5}$ was necessary because we did not have the 3 groups needed to easily see the *two-thirds of*. Each fifth was subdivided into 3 sections. Recall the reasoning outlined here:

$\frac{2}{3}$ of the area

Here is $\frac{1}{5}$. The patterned portion of it shows $\frac{2}{3}$ of the $\frac{1}{5}$, and this reasoning is repeated for each fifth.

In this case (because there was no easier way to model $\frac{2}{3} \times \frac{4}{5}$), we cut the whole (originally divided into 5 pieces to show fifths), into *3 times* as many pieces. This allows us to take $\frac{2}{3}$ *of* each $\frac{1}{5}$. As a result, the total area is now divided into 3 × 5 smaller regions. (Where do these numbers 3 and 5 come from? One side of the area model was divided into 5 equal lengths when the fifths were initially drawn. Subdividing each fifth into 3 pieces divides the other side of the area model into 3. This creates the 3 × 5 space, using the fraction denominators.) Importantly, we note that the patterned region (the answer, which represents $\frac{2}{3}$ of each $\frac{1}{5}$) is a 2 × 4 area (Where do the 2 and 4 come from? The 4 is the number of fifths and the 2 is the number of thirds. The 2 and the 4 are the numerators.) Now we are ready for an exciting observation! Using the model, we find the answer to $\frac{2}{3} \times \frac{4}{5}$ is as follows:

The answer is: a 2 x 4 region (8 units) out of a total region of 3 x 5 (15 units).

A 2 x 4 region is the *numerator* of the answer.

A 3 x 5 region is the total area. There are 3 x 5 pieces in all, so 15 is the *denominator*.

The model shows that

$$\frac{2}{3} \times \frac{4}{5} = \frac{2 \times 4}{3 \times 5}$$

It might be necessary to re-examine the model and the reasoning in the example to connect each stage of the reasoning with the model. Connect each step in the calculation to the model. Think about whether a similar set of values could be represented in the same way, by connecting the location of each number in the fractions with its use in the model. Think about whether a similar model and method could be used if other values were substituted for the 2, 3, 4, and 5. If necessary, create and solve another example.

Reasoning

The example and the solution model were chosen as a generic example of generalizable reasoning. The model suggests an idea or conjecture about a numeric method: we can find the answer to a fraction calculation by multiplying the numerator values to find the numerator, and multiplying the denominator values to find the denominator. Think about how we could use a similar model and reasoning to find the product of any two fractions.

Exploration/Task

Complete the following task to test your understanding. Create a model to find the answer to each question below using reasoning, as in the previous example. As you proceed, think about how you could calculate the answer using the conjecture, outlined above, of multiplying numerators and denominators.

1. $\frac{3}{4} \times \frac{2}{5}$
2. $\frac{5}{6} \times \frac{2}{3}$
3. $2\frac{1}{2} \times \frac{2}{5}$ (Note that this can be completed as two separate calculations, using $2 \times \frac{2}{5}$ plus $\frac{1}{2} \times \frac{2}{5}$. Alternatively, it could also be completed as $\frac{5}{2} \times \frac{2}{5}$.)

Follow-Up and Discussion

Both the model and the conjectured numeric method yield an answer of $\frac{6}{20}$ (or $\frac{3}{10}$) for Problem 1, and $\frac{10}{18}$ (or $\frac{5}{9}$) for Problem 2. What if the calculation involves fractions greater than 1, such as in Problem 3? Using reasoning, your model might have shown that $2\frac{1}{2} \times \frac{2}{5}$ is like 2 of $\frac{2}{5}$ *and* $\frac{1}{2}$ of $\frac{2}{5}$. Since 2 of $\frac{2}{5}$ is $\frac{4}{5}$, and $\frac{1}{2}$ of $\frac{2}{5}$ is $\frac{1}{5}$, we obtain the answer by adding $\frac{4}{5} + \frac{1}{5} = \frac{5}{5}$ (or one whole). Using our newly conjectured numeric method of multiplying numerators and denominators, we also see that $2\frac{1}{2} \times \frac{2}{5} = \frac{5}{2} \times \frac{2}{5} = \frac{(5 \times 2)}{(2 \times 5)} = \frac{10}{10} = 1$. Using both methods, we find that the answer is one whole, or 1. We note that the numeric method actually computed smaller pieces than we really needed, providing the answer in tenths. When we initially found the answer using reasoning, we were able to operate using fifths. While we can always find the answer using the numeric method, it sometimes uses smaller pieces than necessary, and is therefore not always most efficient. Using reasoning as much as possible is not only faster, but is also less prone to computational error. Students should never be penalized for knowing an answer because they are able to visualize it; in other words, if they can visualize a calculation and solve the problem this way, they should not be forced to write out the steps of the numeric method. Because reasoning out the answer is much faster and more reliable than the formal numeric method, it should always be accepted and, in fact, encouraged. Nevertheless, there may be times when the values in the problem are too difficult to work with, and in such cases, the numeric method is very useful.

Practice and Further Exploration

Improve your fluency in the numeric method by using the procedure to solve the following problems. Remember to check for simpler equivalent fractions before you complete the operation. If you are uncertain of your answer, use a model to redo the question.

1. $\frac{7}{8} \times \frac{2}{5}$
2. $\frac{11}{12} \times \frac{3}{4}$
3. $\frac{8}{12} \times \frac{3}{8}$

7.4 Fraction Division: Concept and Contexts

In our work with whole numbers, we explored several different interpretations of division.

Using the *partitive model of division*, we interpret division by a whole number as partitioning, or splitting, an amount into equal groups; for example, $6 \div 2$ can be thought of as splitting 6 into 2 groups (of three). This is also called equal sharing. Division also connects to multiplication; in this case, we see that $6 = 2 \times 3$.

> **CONNECTION**
> To review whole number division models, see section 5.5.

Exploration/Task

Create and use a model that illustrates the partitive (equal sharing) interpretation of division for each of the following problems, and consider the related multiplication statement.

1. $\frac{3}{4} \div 3$
2. $\frac{4}{5} \div 2$
3. $2\frac{1}{2} \div 5$

Follow-Up and Discussion

Problem 1 involves splitting $\frac{3}{4}$ into 3 groups, yielding an answer of $\frac{1}{4}$. In solving Problem 2, we find that the answer is 2 groups of $\frac{2}{5}$. Problem 3 can be solved by thinking of $2\frac{1}{2}$ in terms of halves—5 halves in total. Dividing 5 halves by 5 results in an answer of $\frac{1}{2}$. This can be thought of as sharing the 5 halves equally among 5 people.

Another model of whole number division is the *measurement model*, which requires counting out or measuring out portions using repeated subtraction. It is critically important that students have used the measurement model with whole numbers *before* learning about fraction division, because it is difficult to use the partitive model when dividing *by* a fraction; for example, it is not easy to think of $2 \div \frac{1}{3}$ as splitting the 2 into "$\frac{1}{3}$ of a group." If you are not sure if your students have previously learned the measurement interpretation of division, be sure to provide a few problems that will allow them to practice using this model with whole numbers *before* trying it with fractions.

> **CONNECTION**
>
> To review whole number division models, see section 5.5.

Exploration/Task

Draw a model to represent and solve each problem. In each case, think about the interpretation of the division operation you are modelling.

1. There were 6 metres of rope available, and tie-downs of 2 metres in length were needed. How many tie-downs could be made with the available rope?
2. There were $1\frac{1}{2}$ cups of blueberries in the freezer. If a muffin recipe uses $\frac{1}{2}$ cup of blueberries for each batch of muffins, how many batches of blueberry muffins can be made?

Follow-Up and Discussion

Context is always critically important when making sense of fraction operations, and this is particularly true when exploring division. Drawing a model makes it clear that 3 tie-downs (of 2 metres in length) could be made in Problem 1, and in Problem 2, 3 batches of muffins could be made. In the models of these problems, context makes it clear that we are measuring portions, or counting groups, of the second number.

7.5 Concrete Models and Strategies for Dividing Fractions

Dividing a Fraction by a Whole Number

When dividing an amount by a whole number, even a fractional amount, an equal sharing approach makes sense (see sections 5.5 and 7.4). If students can understand dividing 6 apples among 3 children, for example, it is not that much more difficult to divide $\frac{3}{8}$ of a chocolate bar among 3 children. Making an explicit link to other models of dividing by a whole number can also help students understand this concept.

Exploration/Task

Create and use a model to solve each of the following problems:

1. There are 4 slices of a 6-slice pizza left. If 2 people share the remaining pizza, what fraction of the whole pizza will each get?

2. 3 people ran a relay of $1\frac{1}{2}$ kilometres. How far did each person run?
3. There was $\frac{3}{4}$ of a tank of gas left for a trip to the country. Assuming that the same amount of gas would be used for the trip there and for the trip home, what is the maximum amount of gas that could be used to get to the country?

Follow-Up and Discussion

It is important to continue to provide problems with contexts that require students to use the partitive or equal sharing model of division. Each of the above problems involves this model. In Problem 1, each individual got 2 slices of pizza. In Problem 2, each person ran $\frac{1}{2}$ kilometre. In Problem 3, dividing $\frac{3}{4}$ by 2 requires re-representing the $\frac{3}{4}$ as $\frac{6}{8}$, and dividing this by 2 gives us an answer of $\frac{3}{8}$. In this problem, we also note the connection to multiplying by $\frac{1}{2}$.

Dividing by a Fraction

> **CONNECTION**
> To review whole number division using the measurement or repeated subtraction model, see section 5.5.

> **INSIGHT**
> Division by a fraction is like counting how many of a particular fraction amount we have.

Dividing *by a fraction* can seem more challenging to students, especially if they are not comfortable with the measurement or repeated subtraction model.

What does it mean to divide an amount *by* a fraction? In the previous example of the relay race, we could ask how many runners would each have to run $\frac{1}{2}$ kilometre to reach a total of $1\frac{1}{2}$ kilometres, and a model would show we would need 3 people. An important interpretation of division *by a fraction* involves thinking about *how many sets, groups, or portions* of that fractional amount we have or need.

The counting or measurement (repeated subtraction) interpretation of division by a fraction is the same regardless of the initial quantity being divided.

Exploration/Task

Solve each of the following problems using a model and reasoning:

1. We have 3 litres of paint. We need $\frac{1}{2}$ litre to paint 1 chair. How many chairs can we paint?
2. There is $\frac{3}{4}$ of a pizza left in the school cafeteria at lunch. Each school lunch includes $\frac{1}{4}$ of a pizza. How many lunches can be made using the remaining $\frac{3}{4}$ of a pizza left?

Follow-Up and Discussion

In each of your models, you may have found yourself counting amounts. For Problem 1, if we think about using up 3 litres of paint, $\frac{1}{2}$ litre at a time, we see there are 6 such portions. Similarly, 3 lunches (of size $\frac{1}{4}$ pizza each) can be made from the available amount. Both of these solutions flow directly from the measurement interpretation of division, which uses the repeated subtraction concept.

Each of these examples resulted in a whole number answer—6 chairs painted, 3 lunches made. We now turn to the more complex situation in which the amounts do not work out exactly. In whole number division, the context determined our responses to questions that did not work out exactly. For example, when determining how many buses were needed to transport a group of students, if there were any students remaining after we counted the number of full buses we need to round *up*, regardless of the number of students remaining, or some students would be without transportation. In many contexts, it doesn't make sense to talk about fractional things, such as fractional buses or fractional numbers of children; however, in other contexts, part of a whole has more meaning. For example, knowing we have enough paint to complete half a wall or enough sugar for half a batch of cookies is relevant.

Exploration/Task

Solve the following problem, and include a model in your solution:

> *Crumble topping for a large apple crumble calls for $\frac{3}{4}$ of a cup of flour. Since there is only $\frac{1}{2}$ a cup of flour left, Francine decides to make a slightly smaller crumble. What fraction of the other ingredients should she use?*

Follow-Up and Discussion

Perhaps you drew and thought the following:

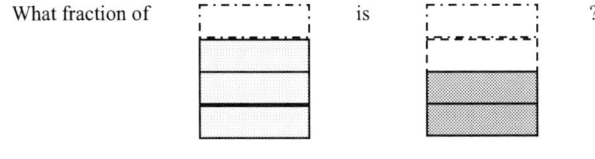

Or, using a different model of a measuring device and slightly reordered reasoning, maybe your model looked like this:

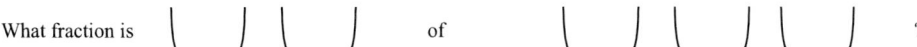

Both models show that we only have $\frac{2}{3}$ of the flour we need. We will need $\frac{2}{3}$ of the other ingredients, too.

Sometimes seemingly simple fraction division calculations are quite challenging to model. This is often the case when the pieces are difficult to compare, and the only way to compare them is to re-represent one or both fractions. Next we will consider the more challenging example of $\frac{1}{3} \div \frac{1}{4}$, which will be solved using two different models.

Example: Using Manipulatives

Using manipulatives, we see that $\frac{1}{3} \div \frac{1}{4}$ can also be shown as $\frac{4}{12} \div \frac{3}{12}$. Using the measurement concept of fraction division, we can understand the question to mean *how many groups of $\frac{3}{12}$ can we make from $\frac{4}{12}$?* (It is so very important for students to be able to think about the operation in a way that gives it meaning. Actually writing down this phrase may be helpful.) Drawing the $\frac{4}{12}$, we look for amounts of $\frac{3}{12}$ in it:

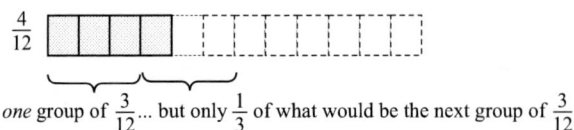

Using this diagram, we see that the answer is $1\frac{1}{3}$. This means that $1\frac{1}{3}$ groups of $\frac{3}{12}$ fit in $\frac{4}{12}$. Using notation, it is expressed as $\frac{1}{3} \div \frac{1}{4} = 1\frac{1}{3}$. We can also use area models to represent and solve such operations.

Example: Area Model

To model $\frac{1}{3} \div \frac{1}{4}$, we first draw thirds, to represent the $\frac{1}{3}$ quantity available:

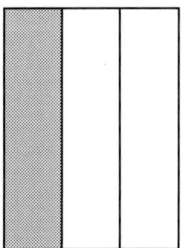

We also need to know how much space is occupied by $\frac{1}{4}$, so that we can count or compare it to the shaded $\frac{1}{3}$. We can draw the fourths *horizontally*:

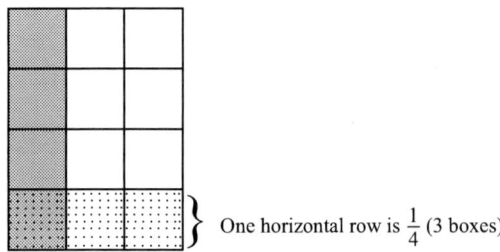

These steps construct 3×4 (12) little boxes. Four of them represent $\frac{1}{3}$, and 3 of them represent $\frac{1}{4}$. Using this model, the question now becomes *how many groups of 3 little boxes are in the shaded $\frac{1}{3}$?* We see:

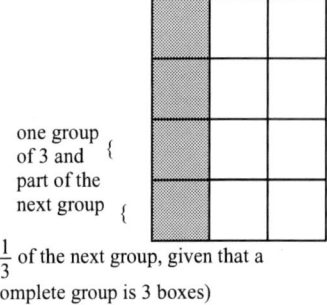

In the shaded region, which is $\frac{1}{3}$, we see one complete group (of $\frac{1}{4}$ or 3 boxes), and $\frac{1}{3}$ of the next group (of $\frac{1}{4}$ or 3 boxes). As we discovered using the first method, $1\frac{1}{3}$ is the answer. The answer implies that there is a complete group of $\frac{1}{4}$, plus $\frac{1}{3}$ of the next group of $\frac{1}{4}$, in the initial $\frac{1}{3}$.

At some point, these models become very cumbersome. It is best to use simpler fractions to practise the concepts. Once students thoroughly understand the concepts, they can start to use the models and reasoning in a way that will allow numeric procedures to emerge.

7.6 Models for Division by a Unit Fraction

> **CONNECTION**
>
> This topic requires an understanding of the concept of division of whole numbers using the measurement model (see section 5.5).

In the previous section, we established the measurement or repeated subtraction interpretation of division by a fraction by showing that division is one way to count how many servings or amounts of a certain size are in a given quantity. For example, dividing an amount by $\frac{1}{2}$ is like asking *how many pieces of size $\frac{1}{2}$ do we have in this amount?*

We have also seen that counting fractional portions using models alone can get messy. Once students understand the concept of modelling, it is time to begin to develop a numeric method. Understanding the effect of dividing by a *unit fraction* is the key conceptual idea needed to develop the formal procedure that is often referred to as *invert and multiply*, hence the importance of this topic.

> **MATHEMATICAL TERM**
>
> A *unit fraction* is a fraction with 1 as the numerator. It represents one piece of the size of the fraction unit specified by the denominator. For example, $\frac{1}{4}$ is one piece of a whole that is made up of 4 one-fourths.

Exploration/Task

Use the measurement model to explore meanings and models of division of a whole number by $\frac{1}{2}$. For example, you might try $3 \div \frac{1}{2}$ or $5 \div \frac{1}{2}$.

Follow-Up and Discussion

Here we look at the example of $3 \div \frac{1}{2}$. Recall that in the measurement model of division, we are counting pieces or parts. Start with 3 one-wholes:

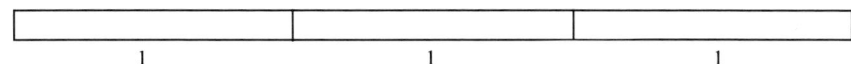

The measurement model interprets division by $\frac{1}{2}$ as *counting the number of one-halves*. To count pieces of size $\frac{1}{2}$, we might draw the following:

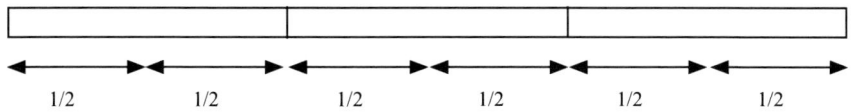

There are 6 pieces of size $\frac{1}{2}$ in 3.

To solve this type of problem, we can also use other manipulatives such as circles. For example, using circles to model $5 \div \frac{1}{2}$ could involve drawing 5 circles and then counting the number of half-circles, which is 10. Note that in this example, when dividing by halves, we ended up with an answer that is *double* the number we started with. Will this always be true? If we divide by $\frac{1}{n}$, will we always get n times as many pieces as we started with?

Exploration/Task

Model another example, such as $2 \div \frac{1}{4}$, and think about the fraction size. Will we end up with 4 times as many pieces as what we started with?

Follow-Up and Discussion

Dividing 2 by $\frac{1}{4}$ is like asking *how many one-fourths are in 2 wholes?* We see 8 of the one-fourths in our model. The idea of four-fourths in each of the 2 wholes is the same as 2×4, or 4 times as many as 2. We also recall that when we divided by $\frac{1}{2}$ we got *twice* as many pieces. Let's explore further to see if this method of counting pieces in one whole will work if the first number is also a fraction.

Exploration/Task

Try these examples to further explore this idea. Model these division statements to see if, when dividing by a fraction such as $\frac{1}{n}$, the result is n times as many pieces.

1. $4 \div \frac{1}{3}$
2. $1\frac{1}{2} \div \frac{1}{4}$
3. $\frac{5}{6} \div \frac{1}{3}$

Follow-Up and Discussion

For Problem 1, you should have found an answer of 4×3 or 12 pieces—3 pieces of size $\frac{1}{3}$ in each of the 4 wholes. For Problem 2, you should have found 6 amounts of size $\frac{1}{4}$. Note that 6 is the same as $1\frac{1}{2} \times 4$ (the 4 is the denominator of $\frac{1}{4}$ and also represents the *number of pieces of size $\frac{1}{4}$ in each whole*). For Problem 3, you should have been able to find $2\frac{1}{2}$ pieces of size $\frac{1}{3}$ in your model of $\frac{5}{6}$. According to our observations so far, this should be the same result as $\frac{5}{6} \times 3$. Multiplying and simplifying shows that $\frac{5}{6} \times 3$ is $\frac{15}{6}$. If we group these pieces in groups of 6 to find the number of wholes in $\frac{15}{6}$, we see that we have $2\frac{3}{6}$ or $2\frac{1}{2}$. This means that there are $2\frac{1}{2}$ pieces of size $\frac{1}{3}$ in $\frac{5}{6}$. We could also use multiplication to check this answer—$2\frac{1}{2}$ pieces of size $\frac{1}{3}$ is indeed $\frac{5}{6}$.

Reasoning

It makes sense that dividing an amount by a unit fraction of the form $\frac{1}{n}$ (remembering that dividing by $\frac{1}{n}$ is like counting the number of $\frac{1}{n}$-sized pieces), gives n times as many pieces. For example, to find how many $\frac{1}{4}$'s there are in 5, we could multiply 5 by 4. We see that instead of dividing by $\frac{1}{n}$, we can simply multiply by n.

Practice and Further Exploration

Create a model to answer each question, and explain your reasoning. Make sure your explanation would make sense to your students. Look for a connection to the previous examples to further verify and make sense of the numeric method just developed for unit fractions.

1. $4 \div \frac{1}{5}$
2. $1\frac{3}{4} \div \frac{1}{2}$
3. $\frac{1}{8} \div \frac{1}{4}$

7.7 Developing Procedures for Dividing Fractions

Many people believe there is only one method to divide fractions numerically—the invert and multiply method. In fact, there are a number of numeric methods that can be used, and we will explore several in this section. In particular, the justification of the traditional (multiply-by-the-reciprocal) method requires understanding dividing by a unit fraction, as was developed in section 7.6.

Same Size Parts

First we will explore the simple case of dividing two fractions with the same denominator, or same size parts.

Exploration/Task

Use a model and reasoning to solve this problem. Be aware of any whole number division you might be doing in your head as you work.

It takes Jessica $\frac{1}{4}$ of an hour to walk her dog around the park. If she has $\frac{3}{4}$ of an hour available, how many trips can she make around the park?

Follow-Up and Discussion

While solving this problem, you may have found yourself thinking, *how many $\frac{1}{4}$ hours are there in $\frac{3}{4}$ hours?* Since the size of the piece being considered is the same—one-fourth—the reasoning behind $\frac{3}{4} \div \frac{1}{4}$ is analogous to $3 \div 1$, because the units (one-fourths) are the same. Try to solve a few more examples using reasoning and models, until you are convinced that fractions with like denominators can be divided by simply dividing the numerators. For example, you might try $\frac{6}{8} \div \frac{2}{8}$. This method of dividing numerators is reasonable and useful. We can also use this method when dividing fractions that can be easily converted to have like denominators.

Example

Converting thirds to twelfths is helpful when completing the following calculation:
$$\frac{2}{3} \div \frac{2}{12} = \frac{8}{12} \div \frac{2}{12} = 8 \div 2 = 4$$

The method of dividing numerators also works even if the division doesn't work out to a whole number. For example, draw a model (or visualize) and use reasoning to verify that $\frac{3}{4} \div \frac{1}{2} = 1\frac{1}{2}$. Compare this to the *dividing numerators* method shown below:
$$\frac{3}{4} \div \frac{1}{2} = \frac{3}{4} \div \frac{2}{4} = 3 \div 2 = \frac{3}{2} \text{ or } 1\frac{1}{2}$$

This solution tells us that $1\frac{1}{2}$ pieces of size $\frac{1}{2}$ are in $\frac{3}{4}$. Can you visualize the model of this problem?

Reasoning

It seems that if we are dividing *pieces of the same size* (fractions with like denominators) we can simply *divide the numerators* to find the answer. This makes sense because the size of the pieces is the same. This is a straightforward and conceptually reasonable method that can also be used to solve division questions that involve unlike fractions that can be easily converted to fractions with the same denominator. It is, however, limited to fractions with the same, or easily convertible denominators; otherwise, it can be cumbersome.

Dividing Numerators and Denominators

Another strategy that is used less often involves dividing *both* numerators and denominators. Since this method is not useful for all fractions, it may be omitted. Justifying it draws heavily on our flexible understanding of order of operations (see section 5.7). In particular, we are reminded that when dividing by a whole number such as 20, we can first divide by 10 and then by 2, or vice versa. For example, if we were dividing by $\frac{3}{4}$, we can divide by $\frac{3}{4}$ by *first* dividing by 3 and *then* dividing by $\frac{1}{4}$. This method also hinges on our willingness to accept multiplication by $\frac{1}{(\frac{1}{4})}$ as equivalent to multiplication by 4. This is reasonable given our work with division by unit fractions in section 7.6. We can think of $\frac{1}{(\frac{1}{4})}$ as $1 \div \frac{1}{4}$, which we know is 4, because there are 4 one-fourths in 1.

The example below uses both of these ideas. It may look somewhat computationally complicated and hence is noted as optional, but all of the concepts used have been discussed.

Example (optional)

Solve $\frac{12}{8} \div \frac{3}{4}$ using *only* what we know about whole number operations, fraction multiplication, and division by unit fractions. Before beginning, model this operation in your mind. If you cannot, draw a sketch. You should be able to see two amounts of size $\frac{3}{4}$ in the $\frac{12}{8}$, to reach an answer of 2. Your mental model will be much simpler if you are able to see $\frac{12}{8}$ as $1\frac{1}{2}$ or $1\frac{2}{4}$, or even $\frac{6}{4}$. It makes locating the 2 groups of $\frac{3}{4}$ much more straightforward. We proceed with the calculation as follows:

$\frac{12}{8} \div \frac{3}{4} = \frac{12}{8} \div 3 \div \frac{1}{4}$ We know we can divide by $\frac{3}{4}$ in 2 stages

$= \frac{12}{8} \div 3 \times 4$ We know division by $\frac{1}{4}$ is like multiplying by 4

$= \frac{12 \div 3}{8} \times 4$ Using the partitive model, we simply divide 12 by 3

$= \frac{12 \div 3}{8} \times \frac{1}{\frac{1}{4}}$ This is tricky, but we know $1 \div \frac{1}{4}$ is the same as 4

$= \frac{12 \div 3}{8 \times \frac{1}{4}}$ We use our fraction multiplication procedure with denominators 8 and $\frac{1}{4}$

$= \frac{12 \div 3}{8 \div 4}$ We know that $\times \frac{1}{4}$ and $\div 4$ are the same

$= \frac{4}{2}$ Here, we divide

$= 2$ And reach our final answer

Using this example to support a generic argument, it seems that separately dividing the values in the numerators and the values in the denominators gives us the same answer as the solution we found using the model, where we found 2 groups of $\frac{3}{4}$ in $\frac{12}{8}$. We already know that a similar idea is true for fraction multiplication—we can multiply two fractions by multiplying the numerators and multiplying the denominators.

The *divide the numerators and divide the denominators* method is sometimes helpful, as in the previous example, but not always, as it can leave a further calculation to be completed, as shown below.

Example

Using the previous method, we begin as follows:
$\frac{12}{8} \div \frac{2}{3} = \frac{12 \div 2}{8 \div 3} = \frac{6}{\left(\frac{8}{3}\right)}$

In this case, we still need to use another method to finish the calculation $\frac{6}{\left(\frac{8}{3}\right)}$.

The Traditional Procedure

Lastly, we turn to developing a procedural method that will work when dividing all fractions. It is important to note, however, that in cases in which a simpler method can be used to successfully solve the problem, the simpler method should be used. The method outlined in this section should be seen as a last resort, used only if the methods described above—visualizing parts, dividing numerators, or dividing both parts of the fraction—do not work out simply. We call this the *chainsaw* method, meaning that it should not be used if scissors will do! This procedure is also a helpful tool that can be used when students progress to algebraic expressions.

Several concepts described earlier in the book are needed to generate our final numeric strategy or method. We will review each of these separately. Although the derivation of the traditional procedure may look complicated, it is in fact just a collection of familiar ideas used together. It is important to remember that we have already established every idea used in the derivation. The full derivation illustrates how these distinct ideas come together to allow us to use the shortcut method, sometimes known as *invert and multiply*. Invert and multiply is simply the end result of performing all of these steps together. After we see how this method works, we won't have to perform all of the steps each time we use it.

For the derivation, we will use the generic example of $\frac{1}{6} \div \frac{2}{3}$. First, we will explore this example with a model, so that we are clear on both the meaning of the problem and the answer. Next, we will review each conceptual idea needed to perform the procedural derivation. Lastly, we will put it all together to illustrate the formal procedural method.

Exploration/Task

Draw a model to represent $\frac{1}{6} \div \frac{2}{3}$. Use reasoning to model the calculation and answer the question.

Follow-Up and Discussion

You may have found that the answer to the question of *how many (or how much) of $\frac{2}{3}$ fits in $\frac{1}{6}$* is that only $\frac{1}{4}$ of the $\frac{2}{3}$ fits in the $\frac{1}{6}$ (because $\frac{2}{3}$ is 4 times as large as the $\frac{1}{6}$). The answer is $\frac{1}{4}$. Numerically, this is expressed as $\frac{1}{6} \div \frac{2}{3} = \frac{1}{4}$.

To follow are the four main conceptual ideas needed for the formal derivation, which then follows. You may wish to refer back to each of the following paragraphs as you go through the formal derivation.

The first conceptual idea is the flexible use of division strategies, which was established when working with whole numbers (see sections 5.5 and 5.7). For example, we can divide by 20 by first dividing by 10, and then dividing by 2 (also noting that $20 = 10 \times 2$). Using the same reasoning, we can divide by $\frac{2}{3}$ by first dividing by $\frac{1}{3}$, and *then* dividing by 2 (observing that $\frac{2}{3} = \frac{1}{3} \times 2$).

The second idea we need to be familiar with is the result of dividing by a unit fraction, which was fully explored in section 7.6. We must be familiar with the idea that dividing by, for example, $\frac{1}{3}$ is like asking *how many pieces of size $\frac{1}{3}$ are there*, and realizing that since there are 3 thirds in each whole, the answer will be the initial amount *times 3*. We see that dividing by one-third, "$\div \frac{1}{3}$," and multiplying by three, "$\times 3$," are numerically equivalent.

The third idea is also related to unit fractions. It is the converse of the previous idea, namely that *division* by a whole number, for example 2, is equivalent to *multiplying* by the related unit fraction, for example $\frac{1}{2}$.

Multiplicative Fraction Representations and Operations   **99**

Lastly, we need to apply the conceptually simple idea that a whole number multiplied by a unit fraction is just the product of the two—for example, $3 \times \frac{1}{2} = \frac{3}{2}$.

Understanding each of these concepts allows us to complete all of the intermediate steps that lead to the final formal fraction division method. In practice, we would like to eventually be able to skip the reasoning and development steps and just use the final result as an accepted procedure. Before we can do so, we need to be sure that the procedure makes sense.

We will work through the derivation example of $\frac{1}{6} \div \frac{2}{3}$. Each step in the method relates to a concept described in the previous paragraphs, which should be reviewed as needed.

Example

$$\frac{1}{6} \div \frac{2}{3}$$

⇓ We know that $\frac{2}{3} = \frac{1}{3} \times 2$ from fraction multiplication

$$= \frac{1}{6} \div (\frac{1}{3} \times 2)$$

⇓ We can divide in two stages, first by $\frac{1}{3}$ and then by 2

$$= \frac{1}{6} \div \frac{1}{3} \div 2$$

⇓ Counting thirds by dividing by $\frac{1}{3}$ is like multiplying by 3

$$= \frac{1}{6} \times 3 \div 2$$

⇓ Dividing by 2 is the same as multiplying by $\frac{1}{2}$

$$= \frac{1}{6} \times 3 \times \frac{1}{2}$$

⇓ We use simple fraction multiplication again

$$= \frac{1}{6} \times \frac{3}{2}$$

To conclude, we see that $\frac{1}{6} \div \frac{2}{3} = \frac{1}{6} \times \frac{3}{2}$.

Reasoning

In the above example, *division by $\frac{2}{3}$* was accomplished by *multiplying by $\frac{3}{2}$*. This method can always be used, by replicating the reasoning described in the example. The fractions $\frac{2}{3}$ and $\frac{3}{2}$ are called *reciprocals*.

Using fraction multiplication, we can finish the division calculation in the example as follows:

$$\frac{1}{6} \div \frac{2}{3} = \frac{1}{6} \times \frac{3}{2} = \frac{3}{12}$$

> **MATHEMATICAL TERM**
> *Reciprocals* are rational numbers with the relationship a/b and b/a, where a and b are not zero.

The final answer to the question is $\frac{3}{12}$. By using manipulatives, or grouping by threes, we can see that $\frac{3}{12}$ is equivalent to the answer of $\frac{1}{4}$, which was found using the model. For at least this one example of fraction division, we notice that the numeric answer found by multiplying by the reciprocal is the same as the answer obtained by using the model.

Exploration/Task

Test the multiply-by-the-reciprocal conjecture for a few more cases. For each problem, solve the calculation using either a model, or an established numeric method such as the common denominator (divide the numerators) method. Next, redo each calculation by multiplying by the reciprocal of the second fraction. Are the answers the same?

1. $\frac{3}{4} \div \frac{5}{8}$ (Note that both a model and the compatible denominators method should yield an answer of $\frac{6}{5}$ or $1\frac{1}{5}$.)
2. $\frac{3}{4} \div \frac{3}{2}$ (Note that both a model and the compatible denominators method should yield an answer of $\frac{1}{2}$.)

Follow-Up and Discussion

In Problem 2, the answer found using a model or the compatible denominators method is $\frac{1}{2}$, while the multiply-by-the-reciprocal method yields an answer of $\frac{6}{12}$. This is an example of the chainsaw method at work—sometimes it leaves an unnecessary mess behind! We may want to express this answer of $\frac{6}{12}$ in its equivalent, and much simpler, form, $\frac{1}{2}$. This is an example of a problem that is easier to solve using a simpler method such as compatible denominators.

We can also derive the multiply-by-the-reciprocal division procedure using a more formal and symbolic approach, which looks more like a formal proof. We caution that such a symbolic derivation might not be helpful to some students, possibly causing some confusion. You may choose to continue working with generic examples, such as those provided above. The argument below is meant to justify more generally that dividing any number A (which could be a whole number, a fraction, or other type of number) by $\frac{n}{d}$ is equivalent to *multiplying* by $\frac{d}{n}$. It might be helpful to examine this formal derivation by comparing each line to the reasoning provided for the equivalent step in the previous example.

Formal Derivation of the Multiply-by-the-Reciprocal Method of Division

From our exploration of flexible whole number division (see chapter 5), we know that

$$A \div \frac{n}{d} = A \div \frac{1}{d} \div n$$

$= A \times d \div n$ There are d pieces of size $\frac{1}{d}$ in an amount

$= A \times d \times \frac{1}{n}$ Dividing by n is like multiplying by $\frac{1}{n}$, e.g., $a \div 2 = a \times \frac{1}{2}$

$= A \times \frac{d}{n}$ Multiplying the last two terms

The method of multiplying by the reciprocal fraction works for dividing by any fraction n/d. We conclude, making the restriction (which is a value that cannot be used) that b, d, and n are not zero, that it is always true that

$\frac{a}{b} \div \frac{n}{d} = \frac{a}{b} \times \frac{d}{n}$

Restrictions will be examined in further detail in chapter 10.

Practice and Further Exploration

Choose a fraction division question that you know you can solve either by modelling or using a numeric method other than the multiply-by-the-reciprocal method, and solve it this way. Solve it once again using the multiply-by-the-reciprocal method. Check to make sure that you get the same answer using both methods.

7.8 Fraction Conversions (Lowest Terms, Improper Fractions, and Mixed Numbers)

In previous sections of this chapter, we suggested that students should not be forced to use simpler equivalent fractions for given fractions before they are ready. In some cases, simpler equivalent fractions are not always clearer or easier to understand. For example, if we are running a 10-kilometre race, it might be clearer to say that we are $\frac{8}{10}$ of the way after running 8 kilometres than to say that we are $\frac{4}{5}$ of the way; however, if requesting a specific amount of fabric at a fabric store, you might be better understood if you ask for $1\frac{1}{2}$ metres of fabric, rather than $\frac{3}{2}$ metres.

MATHEMATICAL TERMS

The fractions from the previous list that are in *lowest terms* are $\frac{3}{2}$ and $1\frac{1}{2}$, which means that they cannot be represented in larger pieces. Examples of *improper fractions* are $\frac{6}{4}$ and $\frac{3}{2}$, which means that they are in the form a/b, but their value is greater than one whole. The *mixed numbers* provided above are $1\frac{2}{4}$ and $1\frac{1}{2}$; each mixed number contains a whole number and a fraction. We also use the term *proper fraction* to represent a fraction with a value that is less than one whole. It may or may not be in lowest terms. For example, both $\frac{3}{4}$ and $\frac{6}{8}$ are proper fractions, but only $\frac{3}{4}$ is in lowest terms.

Example

The example to follow will help us define some new terminology. Consider the following model:

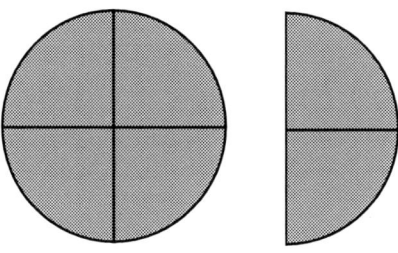

There are many ways to write the number represented by this model. For example, $\frac{6}{4}, \frac{3}{2}, 1\frac{2}{4}$, and $1\frac{1}{2}$ all represent values equivalent to the shaded amount! Can you think of other numbers that represent equivalent values?

Finding Lowest Terms

In the past, formal procedures for changing fraction form were taught, often without understanding. For example, students might have been taught to "cancel" a 2 to change $\frac{6}{8}$ to its lowest terms of $\frac{3}{4}$ as follows:

$$\frac{6^3}{8^4}$$

We strongly caution *against* using this terminology and way of thinking about the method. The terminology can lead to misconceptions and misuse—for example, thinking that $\frac{5}{x+5} = \frac{1}{x}$ because we "cancel" the fives, which is incorrect. If you substitute a value for x, say 10, and check the value of each expression, you will see that the statement is clearly untrue. Instead of using the "cancelling" method, we suggest a more conceptual approach to finding fractions in lowest terms.

Example

When considering the fraction $\frac{6}{8}$, a student might observe:

The model is the same as the model for $\frac{3}{4}$. Pairing up $\frac{1}{8}$ths is like grouping by 2 (or dividing 8 by 2). Groups of $\frac{2}{8}$ are like fourths. The fourths pieces are twice as big so I only need half as many of them (3 instead of 6). I have 3 pieces of $\frac{1}{4}$ ... so I have $\frac{3}{4}$.

> **CONNECTION**
>
> To review factors and how to find them, see section 5.6.

> **MATHEMATICAL TERM**
>
> A *common factor* is a factor of two or more numbers. The *greatest common factor* is the largest of these factors.

Practicing with other examples will allow students to see that grouping both the number of pieces *in one whole* (the denominator value) and the number of pieces there *are* (the numerator value) by the same amount, is like *dividing* each part of the fraction by that number. In other words *dividing* both 6 and 8 by 2 is the numeric version of pairing up or grouping by 2. Numerically, we need a number that divides into *both* the numerator and the denominator; this is called a *common factor*, because it is a factor that both numbers share. The largest number that can be divided into both the numerator and the denominator is called the *greatest common factor*. This number tells us how the parts of the fraction can be grouped.

Below is a neat little diagram that can be used to find the largest factor that two numbers have in common (or the greatest common factor). This technique is sometimes included when teaching about common factors of pairs of whole numbers, but it has more purpose when linked with fractions.

Example

If we want to find the greatest common factor of, for example, 60 and 42, we can factor each one as far as possible. This is called the *prime factorization*; see section 5.6 for more on this.

$60 = 6 \times 10 = 2 \times 3 \times 2 \times 5$
$42 = 2 \times 21 = 2 \times 3 \times 7$

We can now place the factors of each number into a circular model called a Venn diagram, with the *common factors* in the intersection (overlapping area) of the circles:

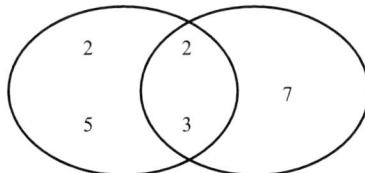

The numbers in the intersection are 2 and 3. These factors are part of each circle; they are factors of both numbers. We multiply 2×3 to get 6, which is the greatest common factor of 60 and 42; of course, both 2 and 3 are also common factors. Simply lining up the factors of 60 and 42 and looking for the numbers that are common to both is another method that can be used to find the greatest common factor.

A working knowledge of basic prime numbers can also be helpful when simplifying fractions. A fraction with a prime number as the denominator is in lowest terms, no matter what its numerator. For example, we can immediately recognize that all fractions that are fifths or sevenths are in lowest terms because no whole numbers will divide the (prime) denominators of 5 and 7. It is important to note, however, that whole numbers can be disguised as fractions, such as $\frac{15}{5}$ or $\frac{28}{7}$.

Example

Let's find out whether the fraction $\frac{46}{437}$ can be simplified. If we knew that 437 was a prime number, then we would also know that $\frac{46}{437}$ is in lowest terms. Or, if we are able to find factors of 46 that are also factors of 437, then we will know that this fraction can be simplified. The factors of 46 are 2 and 23. Finding 23 as a factor of 46 allows us to test to see if 437 is divisible by 23; because 437 is an odd number, we can immediately see that 2 is not a factor. We find that 23 does divide into 437. Using a fraction multiplication procedure developed in section 7.3, we can write

$$\frac{46}{437} = \frac{2 \times 23}{19 \times 23} = \frac{2}{19} \times \frac{23}{23} = \frac{2}{19}$$

We note that in this method, $\frac{23}{23}$ is written as a separate fraction, making it clear that it can be simplified to 1. Again, using this step is more conceptually clear than using rules such as "cancelling."

Although the use of equivalent fractions in lowest terms helps to keep fraction values manageable, this may not always be as important as is often suggested. In many situations, for example, $\frac{2}{6}$ may be just as good an answer as $\frac{1}{3}$; for example, if you know that $\frac{1}{6}$ of a chocolate bar contains 120 calories, it might be easier to think about how many calories are in $\frac{2}{6}$, rather than $\frac{1}{3}$, of the bar. Again, the best form for a final answer often depends on the context.

Exploration/Task

Practice the idea of grouping by a common factor and represent these fractions as stated, and then again in lowest terms. Use the modelling and reasoning provided in the earlier example of $\frac{6}{8}$ and $\frac{3}{4}$, in which the eighths were grouped in twos, to help you find the answers. Use both a model and a numeric method to simplify each.

1. $\frac{4}{6}$
2. $\frac{6}{12}$
3. $\frac{10}{8}$

Follow-Up +and Discussion

For Problem 1, you probably grouped by twos, and found that $\frac{4}{6}$ is the same as $\frac{2}{3}$. For Problem 2, you might have grouped by twos and then by threes, or grouped by sixes to find the solution in one step—either option is fine. For Problem 3, you may have grouped by twos to get an answer of $\frac{5}{4}$, or you might have first represented 8 of the 10 eighths as one whole, which is an example of converting an *improper fraction* to a *mixed number*. You could then work with $1\frac{2}{8}$ to get $1\frac{1}{4}$. Again, either method is acceptable.

Example

Explore representing $\frac{7}{3}$ as a mixed number. $\frac{7}{3}$ can be thought of as *7 pieces of size $\frac{1}{3}$*.

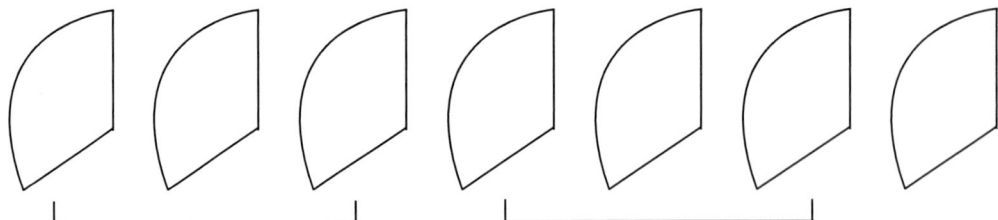

Since we know that three-thirds make up one whole, we can group by threes for each whole. Numerically, grouping by threes is like dividing. Dividing $7 \div 3$ gives us two complete groups of three-thirds and one piece more. Using this method, we can see that $\frac{7}{3}$ is the same as $2\frac{1}{3}$. This is evident in the model, if we rearrange 6 thirds to make 2 wholes.

Exploration/Task

Practice similar reasoning and use a similar model to represent the following improper fractions as stated, and then again as mixed numbers. (As you work, think about how you would re-represent the mixed numbers as improper fractions, using the same reasoning in reverse).

1. $\frac{11}{4}$
2. $\frac{7}{2}$
3. $\frac{12}{8}$

Follow-Up and Discussion

Your models for Problems 1 and 2 should make it evident that the amounts are $2\frac{3}{4}$ and $3\frac{1}{2}$ respectively. You might have any of $1\frac{4}{8}$, $1\frac{2}{4}$, and $1\frac{1}{2}$ as answers to Problem 3. All are correct. Which of these answers is in lowest terms?

Exploration/Task

Challenge yourself to use the same reasoning you used to complete the mixed number conversion above to convert the following mixed numbers to improper fractions. Remember that one whole is four-fourths, so two wholes is eight-fourths, and so on.

1. $2\frac{3}{4}$
2. $3\frac{1}{2}$
3. $1\frac{4}{8}$

Follow-Up and Discussion

For your answers, you should have found the same three values provided in the previous task. The key conceptual element to remember when moving back and forth from mixed numbers to improper fractions is that one whole is four-fourths, or two-halves, or six-sixths, and so on. In real-world contexts, mixed numbers are often the most common and useful form, however, numbers that are going to be used for further calculations are often most usefully left as improper fractions as these may be easiest to further multiply and divide procedurally. In general, exploring multiple numeric forms of fractions, and developing flexibility when working with these various forms, is recommended as best for conceptual understanding, as well as the development of procedural fluency and flexibility. There is not one best way to write an answer; rather, it depends on the context.

Exploration/Task

A student draws a model and uses it to solve $5 \div \frac{2}{3}$ as shown. Is the student's solution correct? If so, explain their reasoning. If not, what is the misconception?

Draw 5 equal parts:

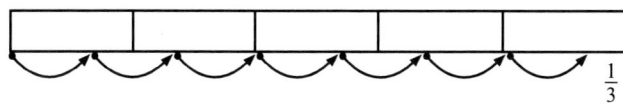

Count two-thirds. There are 7 two-thirds in 5. Since $\frac{1}{3}$ is left, the answer is $7\frac{1}{3}$.

Follow-Up and Discussion

We see the student demonstrating good knowledge of the measurement model with fraction division, and using it correctly to determine the whole number part of the answer. There are indeed 7 complete groups of $\frac{2}{3}$ in 5. It is also true that there is $\frac{1}{3}$ left after all of the complete groups of $\frac{2}{3}$ are counted. However, the remaining $\frac{1}{3}$ is just that: it is the *remainder*, not part of the answer. This remaining $\frac{1}{3}$ represents *one half* of the *next* amount of $\frac{2}{3}$; therefore, the answer is $7\frac{1}{2}$, not $7\frac{1}{3}$. The answer tells us that there are $7\frac{1}{2}$ groups of $\frac{2}{3}$ in 5. The student is well on the way to understanding the division model, but is unclear about what to do with partial groups. Further tasks in which contexts are provided, particularly those chosen to highlight this aspect of the solution process, might be the next step for this student.

Once students have explored fractions conceptually, and developed a selection of methods and procedures for working with them, they should be encouraged to use these flexibly. Forcing students to use a particular method when they prefer another that is equally correct promotes a culture of memorization rather than understanding. Rather, students should be encouraged to share their methods and thinking with each other to broaden their available choices of methods.

Chapter Problems

1. A combined Grade 3/4 class was given $5\frac{1}{2}$ pizzas to share. Two-thirds of the students in the class are in Grade 3. To keep things fair, the Grade 3's will share two-thirds of the total amount of pizza. Use an area model to show the total amount of pizza and what portion of this total amount will be for the Grade 3 students in the class. (Taking it further: Assuming that each pizza has 6 slices, use your model to determine how many total slices the Grade 3 group and how many total slices the Grade 4 group will receive.)
2. Use a diagram to illustrate how many halves there are in $4\frac{1}{4}$. What mathematical operation is being shown in this diagram? Is there more than one possible operation you could use to solve the problem?
3. The students in a Grade 8 class are making centrepieces for their graduation dinner. Each centrepiece includes a decorative bow, made from $1\frac{1}{3}$ metres of gold ribbon. Their teacher has provided them with 12 metres of ribbon to use to make the bows. Use a concrete manipulative or labelled diagram to determine how many bows the students will be able to make with the ribbon provided. (Taking it further: Express this problem symbolically. Link your model to a mathematical procedure that can be used to solve the problem.)

Further Reading

Coughlin, H. A. (2011/2012). Dividing fractions: What is the divisor's role? *Mathematics Teaching in the Middle School, 16*(5), 280–287.

> Coughlin discusses using repeated subtraction when dividing fractions.

Gregg, J., and Gregg, D. U. (2007). Measurement and fair-sharing models for dividing fractions. *Mathematics Teaching in the Middle School, 12*(9), 490–496.

> Gregg and Gregg discuss the mathematics behind division of fractions through the use of word problems to give examples of different conceptions of division and reasoning through dividing fractions.

Izsák, A. (2006). Aspects of mathematical knowledge for teaching fraction multiplication. In S. Alatorre, J. L. Cortina, M. Sáiz, and A. Méndez (eds.), *Proceedings of the 28th Annual Meeting of the North American Chapter of the International Group for the Psychology of Mathematics Education* (pp. 364–370). Merida, Mexico: Universidad Pedagógica Nacional.

> Izsák discusses two Grade 6 teachers and the knowledge they needed as they worked with their classes to develop student understanding of fraction multiplication.

Tobias, J. (2013). Prospective elementary teachers' development of fraction language for defining the whole. *Journal of Mathematics Teacher Education, 16*(2), 85–103.

> This article examines the ways in which prospective elementary teachers develop an understanding of language use for defining the whole, using student work samples and classroom conversations.

Tyminski, A. M., and Dogbey, J. K. (2012). Developing the common denominator fraction division algorithm. *Mathematics Teaching in the Middle School, 18*(4), 248–253.

> Tyminski and Dogbey present a classroom activity that includes activity sheets to develop the algorithm of using a common denominator in division of fractions.

Chapter 8
Decimal and Percent Representations and Operations

8.1 Decimal and Percent: Concept and Contexts

The prevalence of calculators has tended to elevate decimals to the numeric representation of choice for many students. While decimals are certainly important and are often used in real-world contexts, fractions tend to be the representation used in pure mathematics. For example, while a student might record 3.3 as the answer to 10 ÷ 3, a mathematician would likely use $\frac{10}{3}$ or $3\frac{1}{3}$ as the best representation, because it is an exact value. Consequently, a challenge for teachers is to be sure that when a student reaches for a calculator, it is with an understanding of its limitations. In particular, it is critical that students understand decimal approximations of fractions in the case of those that *cannot* be recorded as exact decimals.

Traditionally, decimals were introduced to students after fractions. Students were typically taught how to convert fractions to decimals and decimals to fractions, and the rules to use when applying operations. This chapter focuses on the underlying reasons behind such methods.

Before students begin to think about decimals, it is particularly important that they have a good understanding of place value. Decimals extend the idea of place value from whole numbers, and are based on similar relationships.

Decimals occur naturally in many common contexts, such as currency and distance measurement. For example, websites (such as MapQuest) typically give distances using decimals. Percentages, which are related to decimals, are also frequently used, particularly in the media and retail sales. In spite of this, many people schooled in formal operations with decimals and percentages still struggle to connect these concepts to real-world contexts. For example, people often confuse 0.05% with 5% or 0.05, when in fact 0.05% is actually equal to 0.0005 or $\frac{5}{10,000}$ or

> **CONNECTION**
> To review place value of whole numbers, see section 3.2.

even $\frac{1}{2,000}$. The connection of the values, operations, and concepts to other representations (such as fractions) and to real-world contexts is an important aspect of this chapter.

8.2 Constructing Representations

Decimal concepts can be built using models similar to those used with whole numbers. The fundamental notions of place value for the decimal system naturally extend to decimal values that are not whole numbers. Decimal values with only one digit to the right of the decimal point—such as 0.7—directly connect to fractions with 10 as the denominator, namely tenths. For example, if we represent $\frac{7}{10}$ using fraction strips, our model may look like this

If we then think of base ten blocks, we can choose to represent one whole as

While we tend to represent 1 (that is, one whole) using just one of these small cubes when working with whole numbers, using a unit that can be more easily subdivided helps when representing decimals, thus the choice of the long or rod for one whole.

We know from our experience with whole numbers that moving from 100 to 10 to 1 is like *dividing* by 10—each is *one-tenth* of the previous number. Continuing this 100, 10, 1 pattern demonstrates that it is reasonable to use the notation 0.1 to mean one tenth of one. If one whole is modelled as the diagram above, then one centicube reasonably represents 0.1 or one-tenth.

We can observe an immediate connection to the fraction strip representation. For example, 0.7 or $\frac{7}{10}$ can be represented by 7 one-tenth pieces:

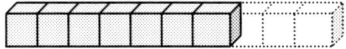

A direct relationship thus exists between tenths and decimal numbers that have one digit to the right of the decimal. This relationship is also true of other decimal places; for example, hundredths and decimal numbers with two digits to the right of the decimal point have a similar relationship, such as $\frac{27}{100}$ and 0.27.

We can also use our existing base ten blocks to model values with several decimal places, as we will explore in the following task. For this activity, you will ideally use a set of base ten blocks. Drawings and a good ability to visualize will suffice if you don't have any handy.

Exploration/Task

Locate the large thousands cube from a set of base ten blocks.

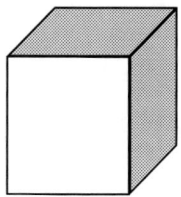

Remember that the thousands cube is like 10 (horizontal) layers, each with 100 (or 10 × 10) cubes. Ten layers of 100 adds up to 1,000. For this activity, we will define the thousands cube as *one whole*. Using the blocks, construct representations of the following values, using the definition of the thousands cube as 1. It might help to think of the thousands cube as 1.000.

1. 0.1 (It might help to think of one-tenth of 1.)
2. 0.3
3. 1.5
4. 0.01 (It might help to think of one one-hundredth of 1, or even one-tenth of 0.1.)
5. 0.123

Follow-Up and Discussion

The model of Problem 1 is one-tenth, and thus the second problem is three-tenths. One-tenth of a thousands cube is a flat, or a 10 × 10 layer of small cubes. Problem 3 should be modelled as one whole and five-tenths. Problem 4 is one one-hundredth, which is a long or 10 small cubes. For Problem 5, we can think of the value, 0.123, as $\frac{1}{10}$ or 0.1, plus $\frac{2}{100}$ or 0.02, plus $\frac{3}{1000}$ or 0.003. Thus the representation of 0.123, when the thousands cube is used as one whole, is one flat, two longs, and three small (one-thousandth) cubes:

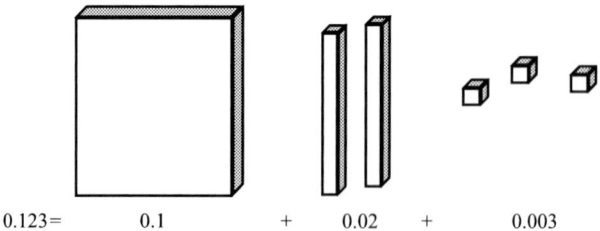

0.123 =   0.1   +   0.02   +   0.003

We can also model percentages using base ten blocks. The term *percent* derives from the French *cent* (hundred)—*percent* is literally *per hundred*. It might also help to think of 100 cents in 1 dollar. *Percentages* are simply fractions of 100. It is even thought that the % symbol might have been derived from the fraction denominator notation /100.

When modelling percentages, it is handy to start with a model of one whole that has 100 subsections. The flat of a set of base ten blocks works well, as does a 10-by-10 or hundreds grid:

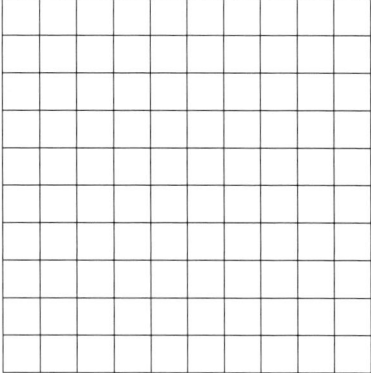

Exploration/Task

Using copies of the hundreds grid, where one 10-by-10 grid represents one whole, model the following percentages. Remember that % means *per hundred* and that there are 100 little squares in the hundreds grid.

1. 2%
2. 99%
3. 0.5% (It might help to ask: *Is this more or less than 1%?*)
4. 150% (It might help to ask: *Is this more or less than 100%?*)

Follow-Up and Discussion

It is easy to see that 2% is 2 small squares on the grid, or $\frac{2}{100}$. It is also fairly simple to see that 99% is 1% or 1 square less than one whole (or 100%). Modelling 0.5% might require comparing it to 1%. Since 0.5% is *half* of 1%, modelling 0.5% on the hundreds grid involves shading half of 1 square out of 100. If we were using the base ten flat instead of the 10-by-10 grid to represent one whole, it would be difficult to show half a centicube. What could we change in our representation of one whole to allow us to model 0.5% with base ten blocks? Modelling 150% requires a complete hundreds square (one whole) plus another 50% (or half) of a square. It might help to think of 150% as 100% plus another 50%.

It is important to interpret percentages as percentages of one whole. Percentages are almost exclusively used to express a percentage of something. For example, at a 30%-off sale you will receive 30% off the original price of an item, and will pay 70% of the original price. In real-world contexts, percentages are generally always used in this way, rather than as values in and of themselves. For example, we might say we bought 0.8 metres of ribbon, but we would not say that we had purchased 80% of a metre. We may, however, say that there is only 50% as much snow as there was at this time last year. An interesting activity for students is to read media reports to examine how decimals and percentages are typically used, and to interpret the values. Many people reading a report that 0.03% of people might be diagnosed with a particular disease think this means $\frac{3}{100}$, rather than correctly understanding it to mean 0.03 out of a 100—or 0.3 out of 1,000—or 3 out of 10,000. It is also surprising how often the media uses percentages incorrectly. For example, a recent ad announced that everything in the store was 60% off, so "you can buy 60% more." Can you find the problem with this reasoning?

As students begin to use the symbolic representation for decimals, they are sometimes confused by place value. They may think that our number system is built on symmetry around the decimal point when actually it is built on symmetry around the *ones place*. This misconception can cause confusion if it is not cleared up early. Consider the following chart:

100	10	1	• 10ths	100ths

If the decimal had a column of its own, the symmetry between the two sides would be lost. While using base ten blocks to model numbers with more decimal places, it is useful to refer to the block or thousands cube as one, the flat or face as one-tenth, the long or rod as one-hundredth, and the small cube as one-thousandth. As students model fractions and decimals, ask them to use a chart like the one below to get used to recording the decimals. This will allow them to become familiar with their symbolic representation. The following example models the value 2.567:

Ones	Tenths	Hundredths	Thousandths
Blocks or thousands cubes	Flats or faces	Longs or rods	Cubes
2	5	6	7

When teaching students to read decimals, it is very useful to read them the same way you would read a fraction. For example, 0.7 should be said as "seven-tenths," rather than "zero point seven" or "zero decimal seven," and 3.42 should be read as "three and forty-two hundredths," rather than "three point forty-two" or "three decimal forty-two." Reading decimals according to tenths, hundredths, and thousandths helps students make meaningful connections to the actual value of the number, and connect to fractional representations.

8.3 Decimal and Percent Addition and Subtraction Concepts and Contexts

Adding and subtracting decimal numbers is based on an understanding of place value, as it is with whole numbers. The location of a value in a number is of fundamental importance. For example, consider the value 7,777.777:

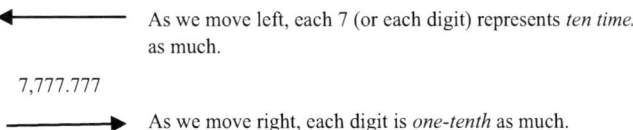

> **INSIGHT**
> Both the digits and their relative position in a number determine the size of the number.

Of course, we might have originally defined the number system differently. The conventional decimal number system is man-made. But, given that we read from left to right, it does make sense that we encounter the digit with the largest value first. For example, in the number 1,900, the 1 (which represents 1,000) indicates a larger amount than the value indicated by the 9 (which represents 900). In other words, as we move to the right in a number, the digits have less and less effect on the size of the number. When determining the amount of a number, we must consider both the location and value of each digit. This is a fundamental concept.

When teaching students to add or subtract decimals, teachers have traditionally focused on lining up the decimal points; however, the concepts we really need students to understand are the framework of place value, how our base ten number system works, and the relevance of positional numeration. As such, the key concept that we must teach with respect to adding or subtracting decimals is that we only operate on like values. In other words, it makes sense to add tens to tens, ones to ones, tenths to tenths, hundredths to hundredths, thousandths to thousandths, and so on. We need to understand the value of the digit given its placement in a number.

8.4 Concrete Models and Strategies for Adding and Subtracting Decimals

All of the models and strategies used to add and subtract whole numbers (see chapter 4) apply to the addition and subtraction of decimals. Base ten blocks and base ten placemats (mats with regions indicating different place value amounts) can be used effectively to model both addition and subtraction of decimals. Before asking students to use these manipulatives, remember to clearly assign values to each piece, including one whole.

Example

Use base ten blocks to model 2.45 + 3.56. Use the block, or thousands cube, to represent one whole. Model 2.45 as follows:

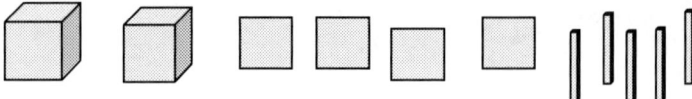

Then model 3.56 as follows:

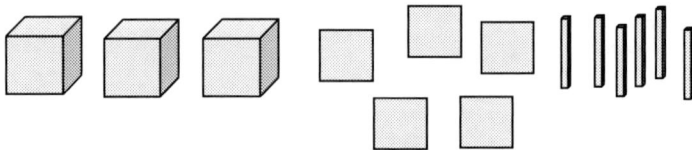

Add the amounts represented in each model:

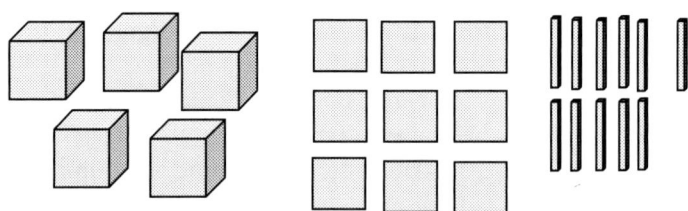

We can see that we now have 11 rods (hundredths), and must regroup. We can use 10 of the rods to form 1 flat (a tenth):

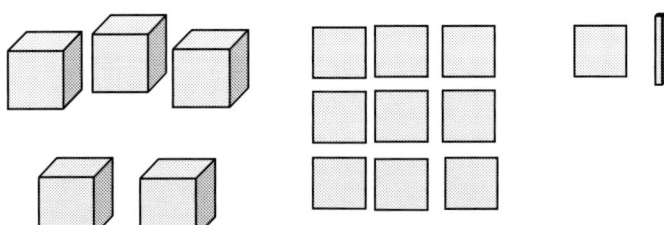

But now we have 10 flats (tenths). We can regroup the 10 flats to form another block (one whole):

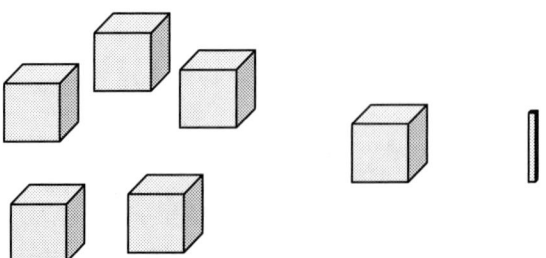

Understanding that each block represents *one*, each flat represents *one-tenth*, and each rod represents *one-hundredth*, allows us to interpret the model to find that the sum is 6.01.

Practice and Further Exploration

Use base ten blocks and placemats to solve the following problems:

1. 1.48 + 4.32
2. 4.42 − 2.37
3. 3.36 − 0.76

8.5 Developing Procedures for Adding and Subtracting Decimals

Just as the models used to add and subtract whole numbers apply to the addition and subtraction of decimals, so do the procedures. Once again, the mathematical concept that must be stressed is that we need to write numbers in a way that allows us to add (or subtract) only like place values.

CONNECTION

To review whole number addition and subtraction operations, see sections 4.2 to 4.5.

8.6 Multiplication and Division Concepts and Contexts, and the Important Role of Estimating

Multiplying decimal numbers can result in products that are more or less than the initial values. The key idea here is whether we are multiplying by a value greater or less than one.

Exploration/Task

Consider the differences between the following problems:

1. Multiplying two decimal numbers that are between natural numbers (i.e., whole numbers greater than zero), for example, 12.5 × 3.85.
2. Multiplying two decimals numbers that are both less than one, for example, 0.5 × 0.75.
3. Multiplying a natural number by a decimal less than one or vice versa, for example, 5 × 0.75 or 0.75 × 5.

How might the products of these questions differ in quantity from one another?

Follow-Up and Discussion

Estimates are the easiest and least error-prone method of determining decimal locations. Working on these problems without having to rely on a memorized rule means that our answers make sense. Unlike a memorized rule, the bonus of truly understanding is that it is difficult to forget this reasoning. When considering Problem 1, we could think *a little more than 12 groups of almost 9*. An appropriate estimate of the answer to this question is 12 × 4, or 48. Problem 2, on the other hand, refers to part of a value less than one. The product will be part of, or smaller than, the second value. Problem 3 suggests a number of groups of something that is less than one, or alternately, part of the whole number.

Each of the previous scenarios could be presented using a different context. Understanding the context allows students to make reasonable estimates that dictate the place value of the answer. Below are some possible contexts for each of the previous problems.

Example

A contextual example for Problem 1 could involve determining the total time it would take a runner to complete 12.5 laps around a track if he averaged a time of 3.85 minutes per lap. What would an appropriate estimate of the total run time be? The context should tell the student that the total time will be very close to 12 (laps) × 4 (minutes), or about 48 or 50 minutes. Regardless of the specific numbers in the estimate, it tells us that the decimal point in the answer should come after the second digit.

A context for Problem 2 could involve determining how much rain fell on a given day if it was 0.5 as much as the 0.75 centimetres that fell on the previous day. By estimating, we can determine that the result will be half of about 0.8 centimetres, or about 0.4 centimetres. Again, the placement of the decimal can be easily predicted after multiplying 5 times 0.75 to get 375. Without any further calculating, we know from our 0.4 rough estimate that the answer must be 0.375.

A question involving a whole number and a fraction less than one, such as Problem 3, could be presented in the context of determining how much pizza is left after a school lunch if each of 5 classes has 0.75 of a pizza remaining. The answer will be more than 1, but less than 5 whole pizzas. If we ignore the decimal and simply multiply the numbers, we again get a product of 375, the same value calculated in the previous context, but since our estimate in this case tells us the product will be more than 1 but less than 5, we know that the answer is 3.75 pizzas.

Similarly, when dividing decimals it is very important to consider context. The measurement model of division provides useful context. For example, we might ask how many lengths of 0.8 metres can be cut from a 6.4-metre length of fabric, or how many days it would take to pave a 54.557-kilometre highway at a rate of 0.89 kilometres per day.

Example

If we consider 54.557 ÷ 0.89, a measurement model could be used to think of counting the number of amounts of just under 1 in about 54. The answer would be greater than 54 (but not greater than 500). A numeric calculation of 54557 ÷ 89 yields 613, so is reasonable to conclude that 61.3 is the answer based on the estimate.

Exploration/Task

Estimate the quotient (answer) for each of the following division questions by considering what the question really means, and estimating the answer. Calculate each answer for the related whole number calculation (for example, 12,324 ÷ 499) using a calculator. Use your estimate as guide when determining where to place the decimal.

1. 12.324 ÷ 4.99
2. 9.672 ÷ 2.872
3. 121.732 ÷ 6.298

Follow-Up and Discussion

When estimating the answers to the above problems, you may have used reasoning similar to the following. The values in Problem 1 are close to 12 divided by 5, so the answer must be between 2 and 3. Problem 2 is estimated as close to 9 or 10 divided by 3, so the answer must be close to 3. For Problem 3, we have something very close to 120 divided by 6, so the answer must be close to 20.

8.7 Developing Procedures for Multiplying and Dividing Decimals

All of the procedures constructed for multiplying and dividing whole numbers, and the connections between multiplication and division of whole numbers (see chapter 5), also apply to multiplication and division of decimals. Although understanding context and content will help students gauge the reasonableness of their answers, it is also important that they are ultimately able to apply the methods described in chapter 5 to decimal numbers. When multiplying and dividing, a good understanding of context, combined with estimating, also helps to ensure that the decimal is correctly placed. Once students have developed these conceptual understandings, they are ready to be introduced to rule-based algorithms, which can be used more quickly. Looking for a pattern with respect to multiplying decimals will help us to generate such a rule.

Multiplying Decimals

What does it mean to multiply one-tenth by one-tenth? If we visualize one-tenth of something, and then using reasoning similar to that used when determining fraction products (determining what one-tenth of one-tenth is), we can imagine slicing the original one-tenth into 10 slices, each of which is 1 one-hundredth; stated numerically, this is $0.1 \times 0.1 = 0.01$.

Let's think about hundredths. Consider multiplying one-hundredth by one-hundredth. Visualize the first one-hundredth of something. This might be one small square in a 10-by-10 grid. Now visualize slicing that one-hundredth into 100 slices—you might mentally blow that one small square up into another 10-by-10 grid. What does each new tiny square represent? If you think about each one-hundredth cut into 100 new tiny squares, this results in 100×100, or 10,000 tiny squares; stated numerically, this is $0.01 \times 0.01 = 0.0001$ or 1 ten-thousandth.

What would happen if we multiplied different decimals, say one-hundredth times one-tenth? Again, visualize 1 one-hundredth of something. Now visualize one-tenth of the one-hundredth. Can you imagine slicing the one-hundredth into 10 slices? What would each slice represent? We find it is 1 one-thousandth, because each one-hundredth would be sliced into 10 slices, resulting in 10 × 100, or 1,000 smaller slices; stated numerically, this is 0.01 × 0.1 = 0.001. Now let's look for patterns. From the three examples above, we see that:

0.1 × 0.1 = 0.01
0.01 × 0.01 = 0.0001
0.01 × 0.1 = 0.001

Rearranging the examples may make the pattern more obvious:

0.1 × 0.1 = 0.01
0.1 × 0.01 = 0.001
0.01 × 0.01 = 0.0001

Do you see the pattern?

It is efficient to recognize that the number of decimal places in the product is equal to the total number of decimal places in both factors. Using this rule allows students to use their choice of strategy or algorithm to do (the number part of) the multiplication, and then to quickly and efficiently place the decimal in its correct position. As with all rules, there is a risk that students may misinterpret or forget this rule. Students should be encouraged to double-check their work using an estimation strategy whenever necessary.

Exploration/Task

Use the number of decimal places rule to determine how many decimal places there would be in the answers to the following questions. You don't need to complete the multiplication; just predict the number of decimal places.

1. 24.31 × 8.3
2. 34.5 × 87.4

Follow-Up and Discussion

If you are unsure of your estimates, you can check the problems using estimation and whole number calculation as we did previously, or you can use a calculator. Next, we will apply the same ideas to division.

Dividing Decimals

With respect to division, we can also use a principle called *compensation* to help us determine the correct placement of the decimal. Consider dividing 13.4 by 3.8. We might think about *how many fours are in almost fourteen*. This estimate tells the logical placement of the decimal; however, what happens when the numbers are larger and less user-friendly? Compensation involves solving a different problem that is constructed to have exactly the same answer as the one we are trying to solve. Let's look at an example.

Example

Consider the division statement $20 \div 4 = 5$. What happens if we multiply both the *dividend* (or numerator) and the *divisor* (or denominator) by the same amount, for example, 3. (See the ideas related to fractions, especially in section 6.7, that remind us that such an operation does not change the final result.)
The result is $(20 \times 3) \div (4 \times 3) = 60 \div 12 = 5$. We see that calculating $60 \div 12$ gives the same answer as $20 \div 4$. Does this always work?

Reasoning

It makes sense that if we take 3 times as much and distribute it into portions that are 3 times as big, we will still get the same *number* of portions (of course, each portion is bigger). What would happen if we multiply both divisor and dividend by 7 in the original $20 \div 4$ problem? We get $(20 \times 7) \div (4 \times 7) = 140 \div 28 = 5$. Think through why this makes sense. In general, as with fractions, we see that multiplying both the divisor and dividend by the same number does not change the value of the final answer. What happens if we *divide* each number by the same value?

> **MATHEMATICAL TERM**
> The *dividend* is the first number in a division statement. In a context, it is the amount to be divided.
>
> **MATHEMATICAL TERM**
> The *divisor* is the second number in a division statement. In a context, it might refer to the number or size of the amounts to be grouped or counted.
>
> **CONNECTION**
> To review fraction operations, see chapters 6 and 7.

Example

Let's try dividing both values in the previous $20 \div 4$ calculation by 2. After completing the division, we obtain $10 \div 2$, or 5, as before.

Reasoning

Again, connecting to fractions, it makes sense that for $20 \div 4$, we can group the amount we have, 20, into groups of 2 (this is the $20 \div 2$), yielding 10 pairs. Using the 10 pairs, we can then count groups of 2 (rather than groups of 4), which is equivalent to $10 \div 2$. If you need to review the measurement model of dividing, see section 5.3.

In general, as with fractions, it seems that as long as we multiply or divide *both* the divisor *and* the dividend by the same number, the quotient remains the same. To apply this method to decimals, we must ask the following question: Is there a number by which we could multiply both the divisor and the dividend that would result in elimination of the decimals? If so, we could simply use whatever algorithm we typically use for division of whole numbers, and know that we are going to get the same answer.

Exploration/Task

1. Consider $12.4 \div 6.2$. What could we multiply both the divisor and the dividend by to construct a whole number problem with the same answer? Change the question appropriately, and then solve it. Note that there may be a simpler equivalent whole number problem.
2. Solve $64.48 \div 5.2$. Begin by multiplying both values by the same number, as suggested in the previous problem (remember that this is referred to as *compensation*), and solve the question, using a calculator if necessary. Now redo the question using estimating and whole number division of $6448 \div 52$, as demonstrated above. Which method is easier? Would you expect students to be able to use the compensation method here?

Follow-Up and Discussion

For the first problem, when using compensation, we find that $12.4 \times 10 = 124$ and $6.2 \times 10 = 62$. We know that $12.4 \div 6.2$ is the same as $124 \div 62$, which equals 2.0; therefore, $12.4 \div 6.2 = 2.0$. Note that $62 \div 31$ is an equivalent problem with the same answer. The second problem, $64.48 \div 5.2$, involves less user-friendly numbers. Using compensation, we can multiply both values by 100: $64.48 \times 100 = 6448$ and $5.2 \times 100 = 520$. We know that $64.48 \div 5.2$ is the same as $6448 \div 520$, which, when divided using a whole number division algorithm, gives us an answer of 12 with a remainder of 208. When using an estimation strategy to solve the problem, we may set up the question as $65 \div 5$, which gives us an answer of 13. Whole number division gives us an answer of 124. Accordingly, the answer to the original problem must be 12.4. What does the remainder of 208, found when we used the compensation method, really mean?

Connecting Remainders and Decimals

The problem $64.48 \div 5.2$, solved above in its equivalent form of $6448 \div 520$, yielded an answer of 12, but with a remainder. It is important to understand the remainder of 208 as part of the next group of 520; hence, the answer found using the compensation method is really $12\frac{208}{520}$. Expressed as a decimal number, this would be a bit less than 12.5 since the fraction part is a bit less than one-half. Since both values in the fraction are divisible by 52, the fraction $\frac{208}{520}$ is equivalent to $\frac{4}{10}$. The decimal equivalent of $\frac{4}{10}$ is 0.4, which gives us a final answer of 12.4. This matches the answer found using estimation and whole number division.

If the fraction $\frac{208}{520}$ could not have been easily simplified into tenths, or hundredths, or another exact decimal, we would be faced with a new problem. In chapter 10, we will further explore values that don't ever resolve to an exact decimal. The fraction $\frac{1}{3}$ is an example. (What happens when we divide 1 by 3? Try this on a calculator.) A calculator is useful for very difficult decimal computations, but the ability to estimate the correct answer remains important; correctly identifying $0.33333...$ as $\frac{1}{3}$ is also important.

While we emphasize the importance of conceptual understanding, we also recognize the benefits of procedural fluency and efficiency. Rules that are based on understanding, such as those we explored above with respect to multiplication and division of decimals, form the foundation of such efficiency.

Practice and Further Exploration

Use one of the whole number algorithms described in chapter 5, along with ideas presented in this section, to solve the following. If possible, solve each problem several different ways. Think about how your students might understand each method you use.

1. 23.4 × 16.7
2. 108 ÷ 43.2
3. 23.04 ÷ 72

Chapter Problems

1. Use base ten blocks to model the following decimal numbers:
   a) 0.26
   b) 0.555
   c) 0.004
   d) 0.997
2. Use base ten blocks, a 10-by-10 grid, or another model of your choice to model the following percentages:
   a) 25%
   b) 5%
   c) 0.5%
   d) 1.25%
3. Tom has read 0.8 of a book that is 550 pages long. How many pages does he still have to read? Use manipulatives or a diagram to model your thinking. If possible, do the problem several ways.
4. Construct a contextual problem for each calculation below, and use the ideas presented in this chapter to solve them. Solve each problem in more than one way, comparing your reasoning and answer.
   a) 21.45 × 0.3
   b) 88.596 ÷ 3.21
   c) 100 ÷ 3
5. A carton of 2% milk says, "37% less fat than whole milk."
   a) Find the percentage of fat in whole milk, according to this statement.
   b) For the same brand of milk, how much more fat (as a percentage) does the whole milk have than the 2% milk? (Hint—it's not 37%!)

Further Reading

Cramer, K. A., Monson, D. S., Wyberg, T., Leavitt, S., and Whitney, S. B. (2009). Models for initial decimal ideas. *Teaching Children Mathematics, 16*(2), 106–136.

   Cramer, Monson, Wyberg, Leavitt, and Whitney explore the use of the 10-by-10 grid to increase students' understanding of decimal representations.

D'Ambrosio, B. S., and Kastberg, S. E. (2012). Building understanding of decimal fractions. *Teaching Children Mathematics, 18*(9), 558–564.

   D'Ambrosio and Kastberg discuss using grids in their work with pre-service teachers to build an understanding of decimals and their relation to fractional parts.

Rathouz, M. M. (2011). Making sense of decimal multiplication. *Mathematics Teaching in the Middle School, 16*(7), 430–437.

   Rathouz relates decimal multiplication to whole numbers, geometry, and measurement.

Chapter 9
Integer Representations and Operations

9.1 Integers: Concept and Contexts

The set of integers is made up of the whole numbers and numbers that are negative, that is, that represent amounts less than zero. Some teachers feel that the study of integers is one of the more problematic content areas in the intermediate mathematics curriculum for students. One reason for this perceived difficulty could be that, historically, the need for and development of negative numbers came about because of algebra, rather than emerging from real-world contexts. For this reason, the classroom use of real-world contexts is challenging and sometimes feels contrived. Having said that, some contextual connections are still helpful. In this chapter, we will strive to balance these two perspectives—algebraic and contextual.

As we have seen in other areas, teaching strategies that rely on the memorization of rules rather than the development of conceptual understanding can be problematic. While quick-fix rules may seem easier to learn and use in the short term, they often generate problems later on, as they are interpreted (or misinterpreted). For example, when teaching multiplication of integers, teachers too often simply train students to *remember the trick* that two like signs result in a positive product, while two unlike signs result in a negative product. While the trick (if remembered and applied in the correct circumstance) will result in the correct answer, it does nothing to instill any understanding of *why* the product is what it is, and typically results in misapplications.

When constructing classroom contexts, care must be taken to prevent the situations from seeming contrived; for example, contexts that require negative numbers can be described using more common terminology, such as a temperature of 2 degrees *below* zero.

> **MATHEMATICAL TERM**
>
> The *integers* are the set of numbers that include the whole numbers together with numbers that move in the opposite direction from zero on the number line, such as −1, −2, and so on.

Exploration/Task

Represent each of the following situations using a number or number statement. There may be more than one way to represent each.

1. Joe has $100 in his bank account. He writes a cheque for $115. Assuming that the bank allows the cheque to clear his account, what is his new bank balance? (Ignore any possible bank charges or fees.)
2. A machine starts to dig a well. Ground level is 25 metres above sea level. The operator digs 40 metres before reaching water. How would you describe the point where water was reached?
3. Elaine has a mystery number for her classmates to guess. The clue she provides is that the number added to 5 gives an answer of 3. What is Elaine's number?

Follow-Up and Discussion

The first example can be represented using whole number subtraction: 100 − 115, and the answer could be written as −15. Alternately, however, we could say, "I have an overdraft of $15," which might be analogous to situations in which we *owe* someone money. The second example is similar—we would likely state the answer as 15 metres *below* sea level rather than using the negative integer −15. The third example, while less contextual, makes it clear that the only number we can add to 5 to get an answer of 3 is −2.

As illustrated, some contexts can be better addressed using subtraction of positive numbers with a qualifier about the answer, than by insisting on the use of negative numbers. Interestingly, research suggests that children who have not yet been formally introduced to integers can still begin to construct the idea of such negative or "minus" numbers as needed in contexts.

9.2 Extending the Number Line

By the time students learn integers in the curriculum, they have usually had many opportunities to use number lines and are generally able to visualize a point on the line that represents zero with whole numbers increasing to the right. When we extend the number line in the opposite direction, to the left, it makes sense to mirror the numbers on the right, creating their integer *opposites*. To represent the integers on a number line, we mark the whole numbers (0, 1, 2, 3,...) and also the same number distances measured in the opposite direction from zero. For example, the value 2 is taken to mean 2 units to the right of zero (or more than zero), and −2 means 2 units to the left of zero (or less than zero).

Temperature is one of the more meaningful contexts for the integer number line and a thermometer is an excellent visual representation of temperatures above, below, and at zero. The thermometer, which is very much like a vertical number line, is a familiar context for integers—at least in northern climates!

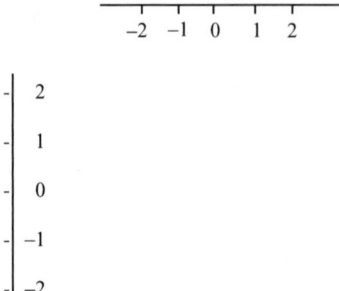

The use of thermometers in earlier grades, and explicitly connecting them to number lines, provides the groundwork for use of the integer number line later in mathematics.

9.3 Exploring Integer Representations, Models, and Manipulatives

In addition to the number line, a useful manipulative to help students understand the concept of integers is a set of counters or "chips" in each of two colours. (Red and yellow are commonly found in sets of classroom manipulatives.) When using coloured counters to represent integers, we need to decide on one colour to represent negative values (here shown as black) and another colour to represent positive values (here shown as grey). It follows then that 2 grey counters would represent + 2:

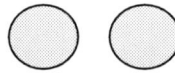

And that three black counters would represent −3:

In the next few sections we will explore how modelling integers with coloured counters can help students better understand integer operations.

9.4 Using Zero Pairs to Represent Integers

The concept of *zero pairs* is necessary for some models of integer operations using counters. Those students who have already experienced alternate representations of numbers may be more willing to accept the idea that the expression 1 + (−1) is zero—and can thus be used as another way to represent zero.

Consider the positive integer +1. We can visualize +1 on a number line by moving one number or unit to the right from zero. This value can also be represented using a coloured counter:

Now think of the negative integer −1. Visualize −1 as a unit distance to the *left* of zero on a number line. This idea—representing a debt of one—can also be represented by a coloured counter:

What happens when we add a "+1" and a "−1" quantity? For whole numbers, we know 1 − 1 = 0. If we imagine a number line, we can see that +1 and −1 are equally far from zero, but in opposite directions. In a sense, they represent a balance of an equal positive and negative distance. Each of the +1 and −1 values can be represented by a counter, giving us a pair. Hence one pair (or many pairs) of positive and negative counters is a way to represent zero. Numerically, +1 + (−1) = 0, or

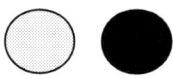

means a net value of *zero*, and we call this a *zero pair*.

> **MATHEMATICAL TERM**
>
> One positive and one negative number of the same value is a *zero pair*; for example, 1 + (−1), which is another way to represent zero. Zero pairs are necessary for some models of integer operations.

Exploration/Task

In section 9.3, we saw that +2 could be represented by two (positive) counters:

What happens if we add a zero pair to our representation of +2? In other words, what integer would be represented by

If we added another zero pair, would the value change?

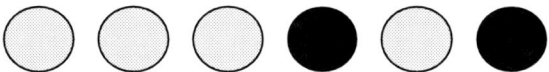

Follow-Up and Discussion

Taken literally, the model of +2 and a zero pair represents 2 + 1 + (−1), so it is equivalent to 2. The next model shows 2 + 1 + (−1) + 1 + (−1); it could also be thought of as +1 + 1 + 1+ (−1) + 1 + (−1). Again, both interpretations show a net or resultant value of 2. We can see that there are many ways to represent any integer using coloured counters.

Practice and Further Exploration

Use coloured counters to represent each of the following integers in at least three different ways:

1. +4
2. −5
3. +10
4. −7

9.5 Integer Addition and Subtraction: Concept and Contexts

By the time students are introduced to integers, they usually realize that addition means combining or bringing quantities together, and that subtraction refers to the inverse operation of addition. Subtraction can also be thought of as the distance between values on a number line, the difference between quantities, or by using the metaphor of removing an amount from a quantity. Although the same ideas apply to adding and subtracting integers, the direction signs can make the operations confusing. Adding or subtracting certain integer values can seem counterintuitive until students develop conceptual understanding. To help students develop such understanding, it is important to provide visual representations and contexts whenever possible.

Historically, integers were defined in response to algebraic situations such as $5 + ? = 3$. In classroom work, we often encounter students who construct the notion of integers themselves, for example, by inventing the following column-wise subtraction method:

$$\begin{array}{r} 302 \\ -109 \\ \hline -7 \\ 200 \end{array}$$

The final step in this method represents integer addition, because students must *add* 200 and (−7) to obtain the final answer. Both the (−7) and the 200 were partial answers—they were the individual answers to the column-wise subtraction, which need to be combined to find the final answer. Flexible use of addition and subtraction strategies may also help here; for example, 200 + (−7) could be solved by writing 190 + 10 + (−7) = 190 + 3 = 193.

As mentioned above, there are a number of potential pitfalls when developing contexts for integer use. It is easy to introduce confusion when real-world contexts in which negative values are used as contextualized positive values, or in which none or zero is taken as the smallest reasonable value. For example, we are much more likely to say, "I have a debt of $5," than "I have negative $5." When we ask a question such as "How much money do I have if I spent all my money and owe Sam $5?" it is not unreasonable for students to answer "none."

Exploration/Task

Explore the appropriateness of the following contexts to prompt integer addition and subtraction statements. If possible, express each situation in more than one way.

1. Lynn has thought of a secret number. When she adds 7 to the number the result is 2. What is the number?
2. Sam gets in an elevator on a floor marked 4 and goes down 6 floors. What might the elevator button read when he gets out? (Assume there is an underground parking garage!)
3. Your credit card statement arrives. At the beginning of the month you had a zero balance and then later there was a charge of $100 to your account. The current balance shown at the bottom of the statement is $60. In addition to those mentioned above, there is one more transaction shown on the statement. What amount is this transaction for? How might this have occurred?

Follow-Up and Discussion

Modelling contexts with both addition and subtraction helps to reinforce the relationship between addition and subtraction. For Problem 1, no context is offered and the only value that accurately solves the problem is −5. Similarly in Problem 3, the transaction, perhaps an item that was returned, must show −$40. However, for Problem 2, depending on how the question is interpreted, one might think of the elevator ride as 4 − 6 or 4 + (−6). The elevator button might read 2 below, or P2, or another variation of this idea.

While the use of context remains important, care must be taken to ensure that such contexts are realistic and not overly contrived. Integers should sometimes be presented strictly in algebraic manipulations, as in Problem 1. At this point, it is suggested that a mix of contextual, algebraic, and other types of representational explorations be used.

Practice and Further Exploration

1. Imagine a game in which a white number cube (die) represents positive values and a red number cube (die) represents negative values. Roll the pair of dice and use the sum of the dice to determine the net value or result of each roll. What are the possible values of each net roll in this game? For classroom use, students may construct their own board game that uses this method of calculating how far a player will move during each turn.
2. A variation on the game would be to allow students to choose whether to add or subtract the two die values, possibly allowing them to do so in either order. How does this change to the rules expand the list of possible values of each net roll?

9.6 Concrete Models and Strategies for Adding and Subtracting Integers

Number lines, temperature models, and coloured counters are useful manipulatives for modelling addition and subtraction and helping students develop an understanding of integer operations. This section will explore each in more detail.

Number Lines and Temperature

To model addition on a number line, we can begin by finding the first value on the line, simply by moving from zero to the location representing that initial amount. Students can be asked: When adding a positive value, would it make more sense to move that many spaces to the right? Or to the left? What about when adding a negative value? Once again, these questions make more sense in a context, such as when representing temperature change. Ask students what happens when you add warmth to a temperature, and they will be quick to tell you that it gets warmer. Ask them what happens when we add something cool, and, again, they will be quick to tell you that the temperature gets colder. Thinking again of our number line as a horizontal thermometer, it follows that adding warmth (getting warmer) requires moving to the right, while adding cool (getting colder) requires moving to the left.

Exploration/Task

1. Use a number line (or thermometer model) to model the following addition statements, as well as their results, using the metaphorical reasoning provided in the previous paragraph.
   a. $(+5) + (+3)$
   b. $(+5) + (-3)$
   c. $(-5) + (+3)$
   d. $(-5) + (-3)$
2. Use the same rationale and models to represent the following subtraction statements and their results. Note that some further reasoning is suggested next to each statement.
   a. $(+5) - (+3)$ If you subtract, or take away warmth, should it get warmer or colder?
   b. $(+5) - (-3)$ If you subtract, or take away cold, should it get warmer or colder?
   c. $(-5) - (+3)$ If you subtract, or take away warmth, should it get warmer or colder?
   d. $(-5) - (-3)$ If you subtract, or take away cold, should it get warmer or cooler?

Follow-Up and Discussion

Asking students to apply the warmth/coolness temperature context to check the reasonableness of their answers when adding and subtracting integers helps them develop conceptual understanding and self-diagnose errors. You can check your understanding of the reasoning hints provided with the previous problems by using a calculator to see if your answers are correct.

Using Coloured Counters to Model Addition and Subtraction

Counters are useful tools to help students understand addition and subtraction of integers, and are frequently used in elementary school textbooks.

Addition of Integers

In section 9.3, we introduced counters as representations of (+1) and (−1). We also determined that the sum of +1 and −1 is always zero. We called this combination a *zero pair*. As with addition models of whole (positive) numbers, adding positives simply results in a sum that is equal to the total of the counters, as shown in the following figure:

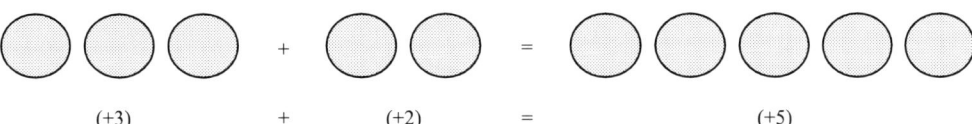

Negative integers can be added using the negative counters. Using the temperature model, we saw −5 + (−3) as starting at minus 5 degrees and adding cold, resulting in the temperature getting colder. Similarly, we see −5 + (−3) as a combination of two groups of negative counters:

In total the 8 chips represent −8. What if the problem involves adding both positive and negative numbers? Again, we can use temperature to provide a meaningful context.

Example

Starting at −3 degrees and then warming up 2 degrees (−3 + 2) can be modelled like this:

But what number does this model represent in total? If we do a little rearranging (which is very easy when working with counters that can be moved around), we recognize zero pairs:

Taken literally, the representation is of −1 + 1 − 1 + 1 − 1. Soon students will learn to immediately recognize the zero pairs, and will see this as 0 + 0 − 1, or even just −1. The answer is −1, which can also be verified using the temperature model discussed previously.

Reasoning

Adding positives and negatives may result in a mixture of positive and negative counters. Because zero pairs can be removed, the result can be simplified to the remaining or net amount shown in the model.

Subtraction of Integers

Subtracting a positive integer value is analogous to subtraction with whole numbers. From the temperature or number line model, we see that questions like 5 − 3 and (−4) − 2 can be understood as starting with the initial value (whether positive or negative), and then decreasing it (by moving down or left) by the units of the second number. Modelling such questions with chips will be examined in the next set of practice questions. Much more problematic is the question of subtracting a negative integer, which will be explored next.

Example

Pretend you are playing Monopoly. You have $5 in your bank, but you owe $2 to your opponent for a fine. Your net worth is $3. Your opponent lands on your space and owes you $2 in rent. The two of you decide to cancel the previous $2 debt to effect this payment. This might be thought of as *removing* your previous $2 debt to your opponent. In terms of operations, we have used subtraction to express the action of removing an amount. The money you now have could be expressed as either $5 − $2 − (−$2) or $3 − (−$2). To check your understanding, model this problem (using play money if necessary). How much money do you have in the end?

Reasoning

At the end of the scenario outlined above, you have the original $5 with nothing owing. We interpret *removing* (or cancelling) the debt of $2 to be like subtracting −2, returning your net worth to $5.

Example

Let's model the previous scenario with the counters. You started with $5, but also had a debt of $2.

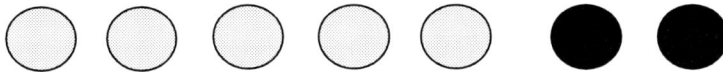

A little rearranging allows us to group the zero pairs, and see that your net worth in the game at that point was $3. There are two zero pairs.

We will now explore the second event in the scenario and model what happens when the $2 debt (the −2) is cancelled or removed. We will interpret the numeric equivalent of removing any amount as subtracting it, so the calculation becomes 3 − (−2). We can interpret the − (−2) as *remove two negative counters*.

Exploration/Task

Given a positive initial value of 3, how can we use the counters model to construct meaning for 3 − (−2), given the interpretation of − (−2) as *remove two negatives*? (Hint: Think about the possible role of zero pairs and alternate representations of the value 3.)

Follow-Up and Discussion

The key idea in modelling examples such as 3 − (−2) when there are no negatives to remove is to re-represent the value 3 by incorporating zero pairs.

After we do so, the action of removing two negatives is straightforward.

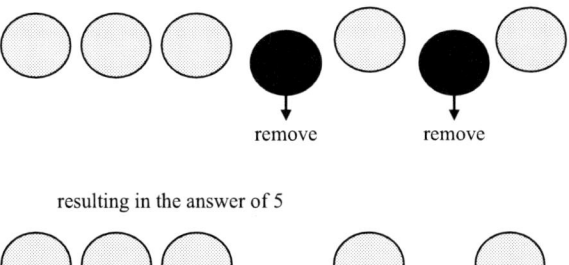

Reasoning

The principle of adding zero pairs to a model can be used anytime we want to remove a certain type of counter (positive or negative) to model subtraction, if that type of counter was not present in the initial model. Adding zero pairs does not change a number's value.

Exploration/Task

Use modelling with coloured counters and reasoning to illustrate and solve 4 − (−3).

Follow-Up and Discussion

We can model the 4 with 4 positive chips, but there are no negative counters in this model, so we can't yet remove 3 negatives.

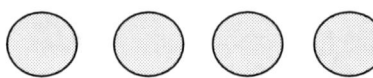

Again, the key idea is to re-represent the value 4 in a way that *will* allow us to remove the −3. Recall that we can add any number of zero pairs to a number without changing its value; therefore, 4 can be shown as

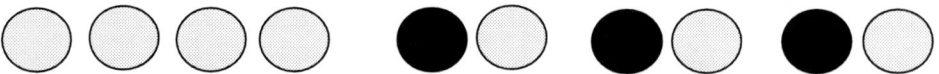

> **INSIGHT**
> Zero pairs can be added at any point to help model an operation.

The re-representation of 4 using zero pairs makes showing the − (−3), or remove −3, operation straightforward. We see that removing 3 negative chips gives an answer of +7.

Two more scenarios remain. Starting with a negative value and then subtracting some of these negatives, as is the case in (−3) − (−2), can be modelled simply by removing some of the negatives. This example results in surprisingly straightforward reasoning, such as: "Start with 3 negatives. Then remove 2 negatives. The answer is −1." In the second case, in which we are trying to remove more negatives than are available, such as in question 4 below, we must again use zero pairs.

Practice and Further Exploration

Use coloured counters to model the following statements. Talk through your reasoning as a student would.

1. (+5) − (+4)
2. (+5) + (−4)
3. (−5) + (+4)
4. (−4) − (−5)

Check your reasoning for questions 2 and 3 by thinking about the temperature change model.

9.7 Practicing Mentally Adding and Subtracting Integers

Students will ultimately develop fluency with integer addition and subtraction, and will be able to reason mentally (or at least use mental images), without the aid of manipulatives. Students are often particularly quick to internalize a directional (number line) approach to the operations, and can be seen using their hands to help them think through a particular calculation.

Practice and Further Exploration

Practice using mental math to increase your fluency with visualized models and images. Solve the following problems without manipulatives or drawn models, instead using mental images of your choice. If possible, do each question in more than one way. If you are unsure, repeat your reasoning using a manipulative or drawn model.

1. $2 + (-4)$
2. $(-4) - 3$
3. $(-5) - (-4)$
4. $(+7) - (+2)$
5. $(-2) - (-3)$
6. $0 - (-2) - (-2) - (-2)$

9.8 Integer Multiplication: Concept and Contexts

By the time the curriculum introduces students to integers, they should have developed a sense of what multiplication means, and should be able to model their understanding in a variety of ways. For example, one interpretation of multiplication is repeated addition; in this case, finding the product of 4×5 means determining how many objects we have in total if we have 4 groups of 5, or $5 + 5 + 5 + 5$. Students should also understand the commutative property of multiplication, which (thus far) we understand to mean that for any whole numbers a and b, $a \times b = b \times a$. This property is fairly straightforward if multiplication is understood with an area model.

As with addition and subtraction, it is difficult to construct real-world contexts for multiplying integers. Integers were not originally developed in response to real-world contexts, but rather as a way of dealing with expressions such as $10 + 2 \times ? = 4$. The following is an example of how we can construct a problem for this calculation.

> **CONNECTION**
> To review whole number models of multiplication, see section 5.2.

Example

I start with $10. I carry out two transactions of the same amount, and am left with $4. What might the two transactions have been? One way to answer the problem is to say, "I spent $3 two times." On the other hand, we might want students to respond that "each transaction was (–$3)." It is important to understand that, in this context, both responses are correct. Algebraically, we will see in the sections to come that (–3) is the correct answer to the equation 10 + 2 × ? = 4, but in words the answer can be described in various ways; saying that we spent $3 is a more conventional way to express the idea in words. An equation is a statement of equality between two expressions that contains an equal sign. Using the coloured counters model, we might show the left side of the equation (the right side having the value 4) as follows:

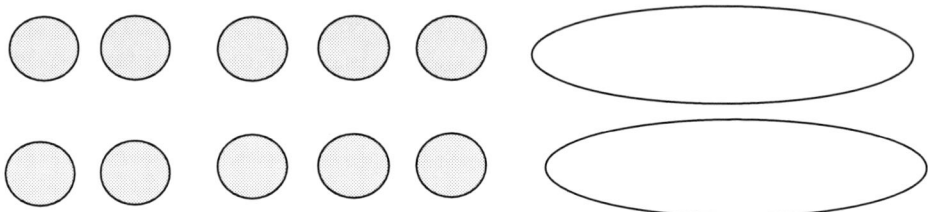

The two ovals are used to represent the two unknown groups.

Exploration/Task

Using the diagram provided, what amount would need to go in to each unknown oval so that the complete model would represent 4? (Remember the idea of zero pairs.)

Follow-Up and Discussion

In the above task, placing 3 negative chips in each oval provides a total of 6 zero pairs. These have net value of zero, so the model now represents 4 as required. With an algebraic interpretation, no possible alternative interpretation of the unknown amount is possible. The answer is clearly defined—only –3 will work to solve the problem when stated as an equation as here.

9.9 Concrete Models and Strategies for Multiplying Integers

As we have in earlier sections of this chapter, we will use contexts where feasible, as well as number lines and coloured counters, to investigate the multiplication of integers. Teachers are encouraged to offer students a selection of models, as some students will prefer one to another. To model multiplication problems in which the first number is positive, it is again helpful to use the repeated addition interpretation. This situation will be explored first, followed by an exploration of problems in which we start with a negative number.

A Positive Number of Groups of a Value

When the first factor in a multiplication statement is positive, the model is relatively straightforward and familiar.

Example

Consider 3 × (−4). Using repeated addition, we can think of this as 3 groups of −4. This can be modelled by constructing 3 groups of 4 negative counters:

This illustrates that 3 × (−4) = (−12).

Reasoning

Using examples similar to the one provided above, it can be seen that a positive number of groups of (either) a positive or a negative number is analogous to repeatedly adding that value. We can also use number lines to model problems in which we begin with a positive value. As we saw previously, to use a number line to model 3 × 4, we can start at 0 and show 3 jumps of 4. The last jump will bring us to 12, which is the product of 3 × 4. We can use a similar model for 3 × (−4).

Example

We determined in section 9.2 that negative numbers can be modelled by moving to the left of the 0 on a number line. To show 3 × (−4) using a number line model, we can reason that we would begin at 0 and make 3 jumps of 4 to the left, ending up at −12, as we found using the coloured counters. This is analogous to 3 temperature drops of 4 degrees each, if the temperature started at 0 degrees.

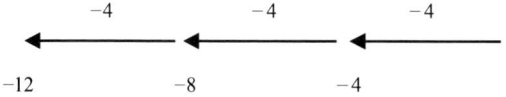

Reasoning

When we are modelling multiplication of a (positive) integer by any other (positive or negative) integer, we can use a number line and simply make jumps to the right or left depending on whether the number is positive or negative. The context of repeated temperature change connects directly with this model.

A Negative Number of Groups of a Value

Negative Times a Positive

Previously, we explored the commutative property for whole numbers, which says that for whole numbers a and b, $a \times b = b \times a$. If we knew that the commutative property also held true for integers, we could simplify an expression such as $(-3) \times 4$ by simply calculating $4 \times (-3)$; as we have already explored this type of question, we know that we can complete this quite easily. Let's examine some other interpretations of a *negative number times a positive number* to see if the commutative property makes sense when working with integer values.

Exploration/Task

Imagine that you have $30 (some of it in $5 bills) in your wallet. Smoothies cost $5. Yesterday, a friend bought you a smoothie, and the day before that, another friend bought you one. You promised to pay them both back later today. How much money do you have in your wallet right now that isn't owed to one of your friends?

Follow-Up and Discussion

If we think about this problem as having $30 plus *owing* two amounts of $5, we might write this as $30 + (-2) \times 5$. This can be argued as contextually subtly different than $30 - 2 \times 5$ because you haven't *yet* removed the two $5 bills. Even so, these two bills aren't really yours—rather, you are carrying a debt, so you have $30 plus a debt of two amounts of $5. According to either interpretation, in the end the net amount you legitimately own is $20. We can see that there is an operational equivalency between $30 + (-2) \times 5$ and $30 - 2 \times 5$. The concept of removing or owing an amount of money provides a useful context for multiplication by a negative number. We could also have recorded the same scenario as $30 + 2 \times (-5)$, which is literally "30 and two debts of 5." The equivalence of these interpretations provides further evidence that the commutative property seems as if it is plausible with integers.

When we previously thought of multiplication as repeated addition, we modelled it by starting at zero and then adding groups. But to model the multiplication of $(-3) \times 2$ using this idea, we would need to add *negative 3* groups of 2. This is difficult to imagine. Coloured counters may help us. Let's go back to an idea introduced earlier, and recall that numbers can be represented in many different ways using zero pairs. We could start with zero—actually a number of zero pairs—and then think of modelling $(-3) \times 2$. We begin with a representation of zero.

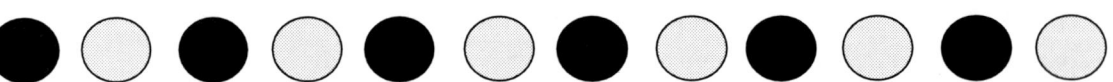

We can use repeated addition to understand 3×2 as adding 3 groups of 2. Evolving from this idea, a helpful interpretation of $(-3) \times 2$ is to *remove* 3 groups of 2.

Exploration/Task

Use this interpretation of *removing* 3 groups of 2 to model (–3) × 2, starting with the counters model of zero shown above. Compare your answer to the answer you obtain for the product of 2 × (–3).

Follow-Up and Discussion

Using the model of zero with a number of zero pairs, we can think of modelling the negative 3 groups by subtracting or *removing* 3 groups of 2 positive counters. After doing this, the model shows 6 unpaired negative counters—a net value of –6. This illustrates that (–3) × 2 = (–6). We also note that had we used the commutative property and changed the question from (–3) × 2 to 2 × (–3), we would have achieved the same result; therefore, we can conclude that the commutative property is plausible with integers.

Practice and Further Exploration

Use coloured counters as illustrated above to solve the following multiplication questions:

1. (–2) × (+5)
2. (–3) × (+7)
3. (–4) × (+2)

Redo each of the above calculations by invoking the commutative property to reverse the order of the two numbers and using a number line model. Reason further about the plausibility of the commutative property in this context.

Negative Times a Negative

Surprisingly, the counters model explored above applies to the *negative times a negative* situation with no further adjustments.

Example

Consider (–3) × (–2) and think *remove 3 groups of 2 negative counters*. Let's begin with a representation of zero:

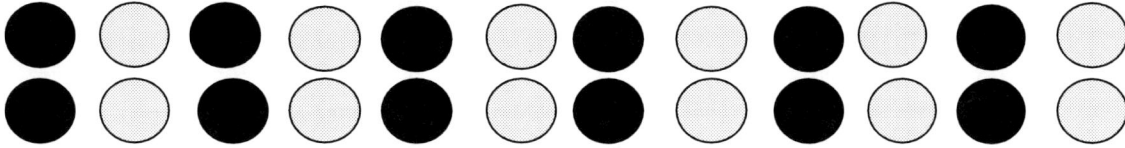

We can use the same reasoning as we did previously to interpret *–3 times* as *remove 3 groups*. This is effectively repeated subtraction. When we remove 3 groups of (–2) from the model of zero shown above, we are left with 6 more positive counters than negative ones—the model shows us that the answer is +6. We conclude that (–3) × (–2) = 6. Construct and solve a few similar examples if you are not yet convinced.

As students become more and more familiar with these examples, they will soon learn to predict exactly how many zero pairs are needed in a model. For example, in the problem just completed, we might have started with 12 zero pairs, but we really only needed 6 of them to solve the problem.

Reasoning

If the first value in a multiplication statement is negative, the repeated subtraction interpretation of *removing* may be helpful. We notice that removing negative amounts from a group of zero pairs results in a surplus of positives. Informally, we can think of decreasing the negatives as a way to make something more positive. After modelling enough examples, students will soon begin to construct and internalize the general procedures for integer products. If they forget these procedures, they have something to fall back on—reasoning!

To follow are two other approaches that we have found are helpful for some students when thinking about integer products. We suggest that students should be provided with a number of methods when learning about integers, as sometimes one particular method will click with a student when another has not. When attempting to model negative groups of any integer on a number line, students can use the knowledge they have of positive groups (that positive groups of positive integers are to the right of zero on the number line and positive groups of negative integers are to the left) to extend their understanding and conjecture that the opposite must be true when taking negative amounts of something. If a negative amount is the opposite of positive amount, it seems likely that when taking negative groups, the product will extend in the other direction. We conjecture that a negative amount of a negative number must extend in the opposite direction to a positive amount of a negative number, which we know extends to the left. Therefore, a negative times a negative must go to the *right* on the number line, because this product must be opposite to a positive times a negative.

Alternately, a negative in front of a value is sometimes used to signify direction, which can be informally modelled by students as they think about walking on a number line. We can think of showing a trip of (−5) as walking backwards 5 steps (from zero) or reversing direction (from facing in the positive direction) and then walking 5 steps. We can interpret (−5) × (−3) as *turn around, and make 5 trips of backwards 3*. Try it! Where do you end up?

This directional interpretation aligns with the idea of vectors in mathematics. For example, we can think about (−1) as a unit vector going in a negative direction. To interpret "− (−1)" we see that the original (−1) vector is *reversed* in direction by the extra negative sign in front, and the final result ends up going in the opposite direction back to positive. What would happen if we kept multiplying by (−1), i.e., continued to add negative signs?

Other students may appreciate a patterning approach. Let's explore a series of products as follows:

3 × (−7) = −21
2 × (−7) = −14   one *less* group of −7 ... or 7 more than the first line
1 × (−7) = −7   again one *less* group of −7
0 × (−7) = 0   one less group of −7 ... or 7 more

Following this reasoning, i.e., that we have one less group of –7, or 7 more, each time, we can continue as follows:

–1 × (–7) = 7
–2 × (–7) = 14
–3 × (–7) = 21

and so on.

Again, we need to think about whether such reasoning will *always* work. All of the models presented in this section align with one another, which provides a strong foundation from which to generalize.

9.10 Developing Procedures for Multiplying Integers

While students are learning to multiply integers concretely, they are developing conceptual understanding. As they build this understanding, they will likely realize that generalizations can be made with respect to the positive or negative value of the product of specific integers. Students should be encouraged to form hypotheses and explore those hypotheses to prove or disprove them, which will allow them to form meaningful generalizations. We want students to not only come to understand, but also to have access to efficient procedures that will help them make calculations quickly and accurately. With sufficient experience, students will generalize on their own.

Multiplying a positive by a negative number is like adding groups of the negative number, so the answer is negative. A negative times a positive number can be thought of as removing positives, so the answer is also negative. Another option is to think of the commutative property for these two situations. Lastly, multiplying a negative by a negative is like removing negatives, so the answer is positive.

Students may eventually generalize that the product of two like-signed integers (two positives or two negatives) is always positive, and that the product of two unlike-signed integers (a negative and a positive) is always negative. Teachers have traditionally relied on these rules without engaging students in activities that provide them with opportunities to develop understanding. We strongly suggest that teachers avoid such practice.

> **INSIGHT**
> An amount that is less negative is thus more positive!

Practice and Further Exploration

Model and solve each problem below. If possible, construct more than one model for each problem.

1. (–4) × 2
2. 4 × (–2)
3. (–4) × (–2)
4. (–1) × (–8)

9.11 Developing Procedures for Dividing Integers

Division by a Positive

CONNECTION
To review whole number division models, see section 5.5.

Using the partitive (equal parts) model of division, division of a positive integer by a positive integer follows fairly seamlessly from whole number division.

We can use similar reasoning to develop the division of a negative number by a positive number.

Exploration/Task

Using coloured counters, use modelling and reasoning to solve the following problems, then write each as a numeric division statement.

1. A $12 prize is shared equally by 4 people.
2. A $12 debt is shared equally by 4 people.

Follow-Up and Discussion

A $12 prize can be thought of as +12, so the first scenario can be represented as 12 ÷ 4. A $12 debt can be thought of as −12, so the second scenario can be represented as (−12) ÷ 4. Using an equal-sharing model of division, in which we split each amount into 4 equal groups, the first answer must be +3 and the second −3. We simply start with whatever type of unit or counter is required, and split it into the specified number of groups. The answer is the value of one of these groups.

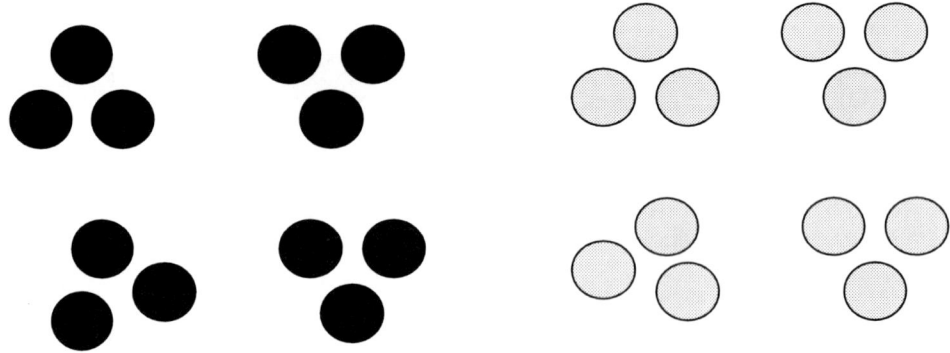

Exploration/Task

Use counters to model and solve (−8) ÷ 2.

Follow-Up and Discussion

In your model, you should have been able to identify 2 equal groups, each of (−4), resulting in an answer of (−4). We also note the connection to multiplication. Two groups of (−4) is (−8), and (−8) split into 2 equal groups results in (−4) in each group. Operationally, we can write these related statements: (−8) ÷ 2 = (−4) and 2 × (−4) = (−8). We are reminded that multiplication and division are inverse operations. The connection between them can also be used to double-check the correctness of our reasoning as we proceed.

Reasoning

Division by a positive number can be interpreted using the partitive (equal sharing or grouping) model of division, in which the first number or quantity in the division statement, whether positive or negative, is split into groups. The answer is the value of each group. In this case, the sign of the answer is the sign of the initial quantity, or first number.

Follow-Up and Practice

Model and solve each of the problems below using coloured counters, then write the related multiplication statement and make sure that it agrees. Note that Problem 1 can be modelled in two different ways—using a measurement or partitive model. Use the same type of model for Problem 1 that you will use to solve Problem 2.

1. 18 ÷ 3
2. (−18) ÷ 3
3. (−10) ÷ 5

Division by a Negative

Once again, we suggest that some contexts and examples used to illustrate operations with negative integers can also be carried out using positive numbers and a description that implies the situation, such as debt or falling temperature. For example, if the temperature changes from 8 degrees to 3 degrees, we might say that the temperature *has fallen* 5 degrees rather than saying the change is −5. When working with negative numbers, the connections to the real world become more tenuous, and discussing this with students, rather than forcing them to contrive situations using negative integers, can be helpful. Purchasing and banking provide possible contexts, although they may not be fully within the realm of students' experience, so some explanation may be required.

Like Signs

Imagine your bank balance is zero, but you are allowed a certain overdraft. You make a series of $4 purchases, after which your bank balance is −$12. How many such purchases were made? Using the measurement model of division, we can think *how many −4s are in −12?* This is analogous to (−12) ÷ (−4). But we can also think of amounts owing or spent, and use 12 ÷ 4.

CONNECTION

To review the measurement model of division, see section 5.5.

Exploration/Task

Model (−8) ÷ (−2) with counters and solve. Write the related multiplication statement and check that the division and multiplication statements agree.

Follow-Up and Discussion

Because the signs are the same, the measurement model of division is helpful; we are counting or comparing things that are alike. We can make a group of 8 negative counters and determine how many groups of 2 negative counters we can make out of these 8 negatives. We see that (−8) ÷ (−2) = 4, i.e., there are 4 groups of (−2) in (−8). This agrees with the related product 4 × (−2) = (−8).

Reasoning

We already know that a positive divided by a positive is positive. When dividing by like signs, even if both values are negative, we can use the measurement model of division. Like signs implies that what we have to divide, and what we are counting by, are the same type of unit. Because we are counting quantities of the same sign, the result will be positive. We can think *how many of these in this*. If the signs are the same, the answer will be a positive number of groups.

Unlike Signs: A Positive Divided by a Negative

Division of a positive number by a negative number is the most difficult combination to connect to a realistic context, or to model concretely. One way around this challenge is to avoid the issue by making use of ideas we are already familiar with as a way to solve such problems. For example, we know from working with fractions that 6/3 is the result of calculating 6 × 1/3. It makes sense that 6/(−3) = 6 × (−1/3), which might be solved by combining the ideas of fraction and integer multiplication. In this case, we might conjecture that the answer will be negative.

Let's explore further with counters, using the example 8 ÷ (−2). Neither of our division models—measurement or partitive—is very helpful in this case. Neither "how many (−2)s are in 8?" nor "what is 8 split into (−2) groups?" seem to make much sense. We will explore this example further with the counter model. When we look at 8 counters, we can't find any (−2)s:

Some children might (quite reasonably) argue that the answer to 8 ÷ (−2) must therefore be zero, by reasoning that there are no groups of (−2) in 8. Certainly there is no negative number of groups.

Looking further at the model of 8, we do see 4 groups of 2. Perhaps we might predict that the answer for groups of (−2) would have to be the *opposite* number of groups, so the answer might be (−4); however, *why* this makes sense is still unclear.

As suggested earlier, another option might be to use ideas that are already familiar. For example, since we know that 8 ÷ (−2) is the same as 8 × (−1/2), which has product of (−4), we can predict that the answer will be a negative (−4), assuming that we are willing to combine what we know about fractions and integer multiplication.

Yet another interpretation uses the number line model. We might ask, "starting at zero, how many jumps of (−2) … (which is 2 to the left) … take us to 8?" Again, initially the answer seems to be "none." The reasoning is a bit abstract this time, but still manageable. If we take one jump of (−2) in our attempt to get to +8, we are clearly going the wrong way. We get even farther from 8 if we take another such jump. We need to consider going the *opposite* way. We can think of this as a reverse or negative jump. In total, we might need to take 4 reverse or backwards jumps, each in the opposite direction to one jump of (−2). This reasoning yields the idea of the opposite of 4 jumps, or 4 *backwards* jumps, which might make the answer of (−4) plausible.

It is a good idea to have students try this on a life-size number line drawn in chalk outside on the playground, or marked on the floor using tape. If a student makes repeated jumps of (−2) from zero, they will get farther and farther from 8, but if they make these jumps the *other* way—in a backwards or "negative" direction—they will reach 8.

Still another interpretation, which uses our knowledge of the relationship between multiplication and division, is to explore the inverse statement of 8 ÷ (−2). If we are trying to solve (+8) ÷ (−2) = ?, then we know that (−2) × ? has to be 8. Again, the only possible answer is (−4), which agrees with the previous conjectures.

More formally showing a measurement (or repeated subtraction) interpretation using coloured counters does work, but students might require a bit of guidance, as this model has to be thought of a bit differently. We will start by reviewing the parallel reasoning using positive values. We recall that when using the whole number repeated subtraction interpretation of division, for a question like 8 ÷ 2, we start with the first number, 8, and then remove or subtract groups of the second number, 2, until we reach 0. The answer is the number of subtractions, which tells us the number of groups. We might write: 8 − (2) − (2) − (2) − (2) = 0. There are four 2s in 8, so the answer is 4.

We could have also written this as a repeated addition statement, and reached the same conclusion. We might ask: "starting with 0, how many twos do we need to *add* to reach 8?" We might write: 0 + (2) + (2) + (2) + (2) = 8. We see that there are 4 twos in 8, so the answer is again 4. Since multiplication and division are inverse operations, these two methods work together, and give us the same answer. We are now ready to explore 8 ÷ (−2).

Example

Model 8 ÷ (−2) using the repeated subtraction/repeated addition reasoning illustrated above. The repeated addition version of the reasoning is easier to interpret with integers. We ask: "starting with 0, how many (−2)s do we need to add to get to 8?" Let's start with one group of (−2):

Since we are trying to get to 8, not −8, we appear to be heading farther from our target, rather than towards it. If we added 4 groups of −2, we would reach −8, not 8. At this point you might ask students, "what can we do?" The idea is to head the *other* way, by *removing* a group of (−2) from 0, which we know how to do from our work with zero pairs. We will think of removing a group, and (−1) of a group, as analogous. Again starting from 0,

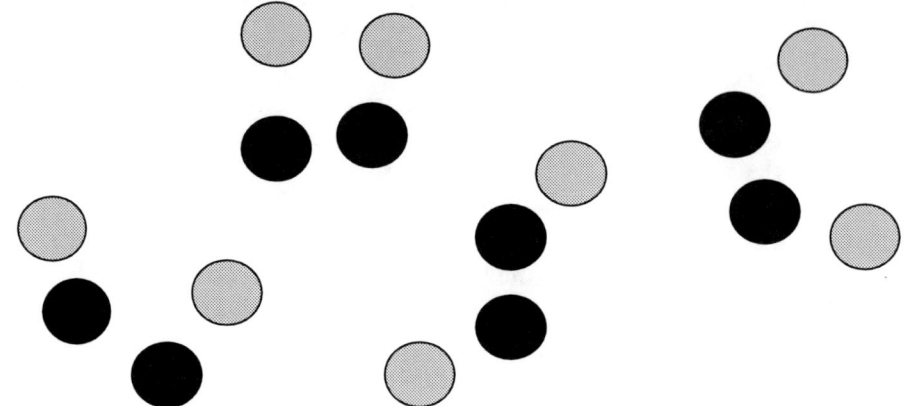

but this time using a zero pairs representation of 0, we now try *removing* groups of −2, with each removal thought of as (−1) of (−2) or (−1) × (−2).

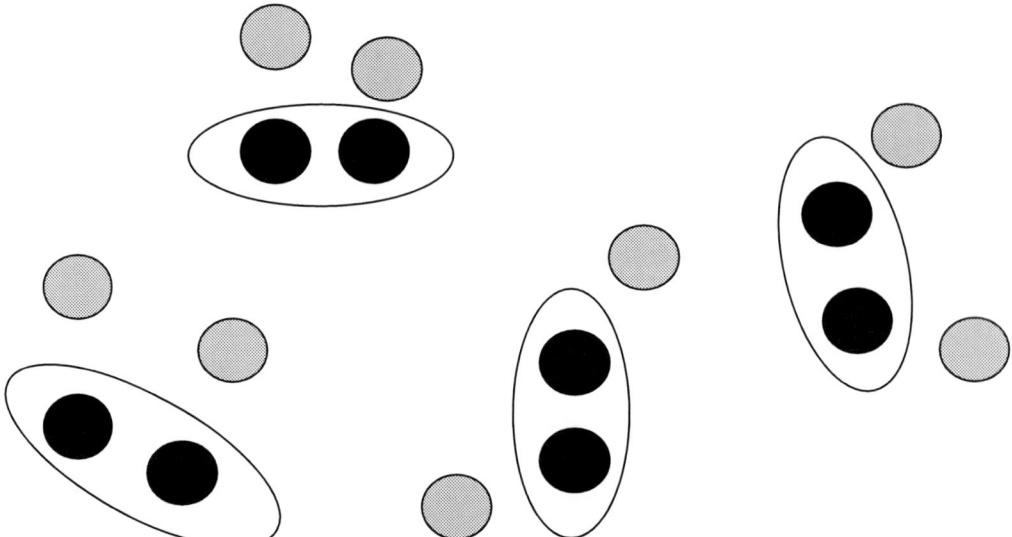

After doing this 4 times, described by (−4) × (−2), we see the result of 8, as desired. The model tells us that to reach 8 from 0 using groups of (−2), we need (−4) such groups.

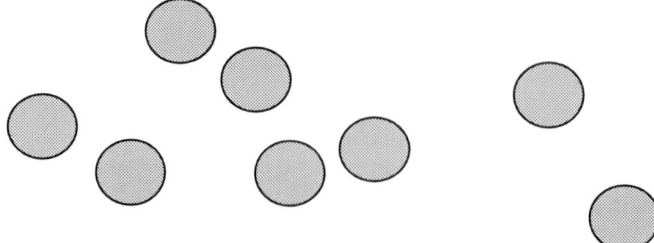

The answer to the question, "how many groups of (−2) in 8?" is (−4). We note that this agrees with the answers found using all of the previous suggested methods.

Reasoning

By using multiple lines of reasoning, and drawing heavily on the relationship between multiplication and division, we conclude that a positive integer divided by a negative integer yields a negative integer.

Exploration/Task

Express the following using as many expressions as you can, which could use positive or negative values. After generating the expressions, look for the equivalence among them.

Thomas arrives at a community yard sale with a few items to sell, and no money in his pocket. He leaves with $20. At the yard sale he makes a number of $5 transactions. How many $5 items did he purchase?

Follow-Up and Discussion

This scenario can be expressed by defining either the money in Thomas's pocket, or the purchases made, as positive, and students shouldn't be forced to think of it in one way or the other. However, if we *compare* students' different methods, we might have a nice set of equivalent expressions to use for discussion. Student answers may include the following:

1. $20 \div 5 = 4$. Perhaps he sold 4 items for $5 each to gain $20.
2. $20 \div (-5) = (-4)$. This is like asking "how many times did he spend $5 to gain $20?" He did the opposite of spending $5 amounts, and he must have sold 4 items of $5.
3. He started at zero. The effect on his money of n $5 purchases is $n \times (-5)$. He ended at $20. So $0 + n \times (-5) = 20$, therefore, n is (-4). He made (-4) purchases of $5. Maybe he sold 4 items.

While this last situation (of a positive divided by a negative) is the most challenging variation of integer division, we can still make sense of it as illustrated. The connection to multiplication here is deeply important as a key contributor to the reasoning.

Reasoning

After gaining experience, students will begin to generalize the results of these operations. They might note that the quotient of two like-signed integers is always positive, because we are counting things that are the same. A negative divided by a positive is negative, as we are simply counting the number of negatives in each new group. In fact, the quotient of two unlike-signed integers is always negative no matter where the signs are located. Students should be allowed to come to these generalizations themselves, without being told. The beauty of mathematics is that students, when allowed enough time and exploration, will indeed do so.

Practice and Further Exploration

1. In each case, construct a model that aligns with the calculation and use it to solve the problem. Try to create more than one model if possible. Write the related multiplication statement for each and check that it agrees.
   a. $15 \div 3$
   b. $(-15) \div 5$
   c. $(-15) \div (-5)$
   d. $15 \div (-5)$

e. $10 \div -2$
   f. $12 \div -3$
2. Build fluency by redoing each calculation in Problem 1 using a method based on mental reasoning. Double-check that each answer aligns with the models you created for Problem 1.

Chapter Problems

1. Choose two single-digit negative numbers. Using a concrete manipulative or diagram of your choice, model why the product of your two numbers is positive, and think about the reasoning that students might use to understand this idea.
2. Use the same two single-digit numbers, but this time assign one of them a positive value. Using a concrete manipulative or diagram of your choice, model why the product of your two numbers is now negative. Think about what reasoning students might use to understand this.
3. Using a concrete manipulative or diagram of your choice, create a generic model that shows the following general statement to be either true or false: Subtracting a negative value from any positive or negative integer is equivalent to adding its opposite (positive) value to the same integer.

Further Reading

Andrews, D. R. (2011). Integer operations using a whiteboard. *Mathematics Teaching in the Middle School, 16*(8), 474–479.

> Andrews describes a classroom lesson that uses an interactive whiteboard to model integer operations in order to build a discussion of technology in the classroom.

Lamb, L. L., Bishop, J. P., Philipp, R. A., Schappelle, B. P., Whitacre, I., and Lewis, M. (2012). Developing symbol sense for the minus sign. *Mathematics Teaching in the Middle School, 18*(1), 5–9.

> The authors discuss the three definitions of a minus sign and their implications for classroom use, as well as ways to support students in developing these definitions.

Sarina, V. (2012). Minus times minus equals plus. *Ontario Mathematics Gazette, 50*(3), 37–39.

> Sarina begins with a discussion of the history of negative numbers and then discusses the use of the "Computing Gnome" on a number line to calculate operations using a concrete method.

Chapter 10
Beyond Integers

10.1 Numbers on the Number Line

Children naturally begin their exploration of numbers with the counting, or positive whole, numbers. These numbers are often represented on a number line to show their relative value and as a model for learning. Number lines are typically used informally at the elementary level, and often it is not specified which type (or set) of numbers—whole numbers, fractions, integers, and so on—is represented on a particular number line. Number lines that represent only whole numbers or integers and do not show any fractions between them really just indicate separate locations or *points* on the line, because there are no numbers defined between points. While we typically illustrate an integer number line like this,

we are actually implying a set of points (with no values between them) like this:

Nevertheless, the line is usually drawn, and generally referred to as a number line.

Once students start learning about fractions and decimals, it becomes necessary to show the parts of the number line between the whole number (or integer) values. An *open number line* uses only labelled values as needed. For example, we might get a rough idea of the relative sizes of 1.4 and 5/8 by showing them on a number line with whole number values that we know, such as 0, 1, and 2, which are often called *benchmarks*.

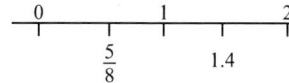

Estimating numbers using rounding can also be made easier by using a number line model. When examining the model above, we notice that 1.4 is closer to 1 than it is to 2 on the number line; therefore, if we wanted to round 1.4 to the closest whole number value, we would round it down to 1. In contrast, 1.7 is closer to 2, so it would be rounded up to 2. By convention, a value right in the middle is rounded up, so if 1.5 is to be rounded to the nearest whole number, it is also rounded to 2. Similarly, 150 rounded to the nearest hundred is 200. We can always find the number between any two decimal values by adding them and dividing by 2. This process can, theoretically, be repeated as many times as desired to "fill in" the number line.

> **MATHEMATICAL TERM**
>
> In *continuous* sets of numbers there are no gaps or spaces between the values on a number line.

> **MATHEMATICAL TERM**
>
> Whole numbers and integers are *discrete*, which means that there are spaces or gaps between them on a number line.

> **CONNECTION**
>
> To explore discrete and continuous data in the context of data management, see section 15.2.

10.2 Density: The Concept of "Between"

As discussed in the previous section, if we want our number line to include all of the decimals and fractions between the whole numbers, we can draw a line between any whole number or integer points. Mathematicians might call this number line the *real number line*, and the group or set of numbers that fills in this line the *real numbers*. The real number line includes values as close to any given value as we want, and always has a value between any two given points. It includes strange decimal numbers such as the one we call *pi* (which begins 3.1415, and continues infinitely with no pattern, which we will explore in chapter 14). A simple way to find the number between, for example, 5.12345 and 5.12346, is to calculate a simple average by dividing the sum of the numbers by 2: (5.12345 + 5.12346)/2 = 5.123455. This will be examined in greater detail in chapter 15.

Because we can always use this method to find the value between two numbers, we can reason that there must be an infinite number of values on the real number line; mathematicians use the word *dense* to describe this. The real number line represents *continuous* data—with no spaces between values—while whole numbers and integers are *discrete*, which means that there are spaces or gaps between them on a number line.

10.3 Formal Descriptions

Mathematicians refer to the different classifications or types of numbers described in section 10.2 as *sets*. They use a prescribed formal notation to specify that a value is from a particular set or classification of numbers. For example, if an unknown value represented by the variable x could be any number in the set of real numbers, this can be described as follows:

$$\{x \mid x \in R\}$$

This symbolic notation can be read aloud as "a variable x, such that x is an element of the set of real numbers." If there was a particular value that x was not allowed to assume, this might also be specified. For example, if x could not be zero, this might be written as follows:

$$\{x \mid x \in R, x \neq 0\}$$

Can you think of a situation in which we might want to remove zero from the allowable values? Below, we will explore what happens if we divide an amount by a smaller and smaller number.

Exploration/Task

Pick any whole number, and try dividing it by increasingly smaller values. For example, you might try dividing 10 in the following ways:

10 ÷ 0.1
10 ÷ 0.01
10 ÷ 0.001

$10 \div 0.0001$
$10 \div 0.00001$

What happens as you make these calculations? What if you continued to divide 10 by a smaller and smaller number (that is closer and closer to zero)?

Follow-Up and Discussion

As we divide by increasingly smaller numbers, we find a pattern—the result gets larger and larger. What happens when we attempt to divide by *exactly* zero? One way to think about this is to look at the effect that dividing by exactly zero has on multiplication. We know that multiplication and division are related, and that, for example, $2 \times 3 = 6$ means it is also true that $6 \div 3 = 2$. If it were possible to have $6 \div 0$, then there would have to be a number for which $0 \times ? = 6$. Since there isn't—zero multiplied by anything is zero—we say that division by zero is undefined. Zero must be excluded from the set of allowable denominator values for a division calculation—this is called a restriction. Such a set of allowable values for a variable also has a mathematical name—it is called the *domain*.

The domain for the variable in the $10 \div x$ calculation illustrated above could be, for example, all of the integers, or even all of the real numbers, as long as zero is omitted in both cases. If the value was to be taken from the set of integers, with zero excluded, we would write:

$$\{x \mid x \in I, x \neq 0\}$$

Another term related to the idea of domain is *range*. This refers to the set of all resultant values of a calculation or expression.

The range is often simply the same set as the domain. Here is an example of a relation for which the range will be only positive values, or possibly zero.

$$y = \sqrt{(x^2)}$$

If the domain is the whole numbers, sometimes written with a W,

$$\{x \mid x \in W\}$$

then the range would be the same—it would also be W. But if the domain was the set of integers,

$$\{x \mid x \in I\}$$

then the range for

$$\sqrt{(x^2)}$$

would be only the positive integers, written as

$$\{y \mid y \in I, y \geq 0\}$$

Alternately, we see that the range for this example, the positive integers including zero, is just the same as the set of whole numbers, which gives us another way to write it. These terms can be confusing for students, and we suggest that they not be introduced until they are really needed; however, it is necessary for students to recognize relatively early on that division by zero is undefined. At the earliest stages, it might be sufficient to simply state that for a given denominator value d, $d \neq 0$.

> **MATHEMATICAL TERM**
>
> The *domain* is the set or list of allowable values that can be used as unknowns or variables in a given expression. For example, the domain might be restricted to whole numbers (and thus not include fractions or decimals) for a given context.

> **MATHEMATICAL TERM**
>
> The *range* of a calculation or expression is the set of numbers of resultant values.

> **MATHEMATICAL TERM**
>
> A *rational number* is one that can be written in the form *a/b*, where *a* and *b* are integers and *b* is not zero. An *irrational number* is a number that cannot be written in this manner.

10.4 Irrational Numbers

Students may enjoy the term *irrational*, which came about in contrast to the rational numbers, which are those that can be expressed as fractions, or in fractional form. Irrational numbers are decimal numbers that cannot be expressed in fractional form. The number we call pi or π is an example of an irrational number.

Exploration/Task

Examine the decimal equivalents of the following fractions by entering them into a calculator as a division expression. For example, examine $\frac{2}{9}$ by entering 2 ÷ 9.

1. $\frac{2}{9}$
2. $\frac{1}{3}$
3. $\frac{10}{11}$

Follow-Up and Discussion

For each of the above problems, you would have found a decimal expression that seemed to go on and on. While you might initially assume that a value such as 0.9090909… is irrational (as it doesn't look like a number that could be easily written as a fraction), students might find to their surprise that it *is* a fraction, discovered when they divide 10 ÷ 11. Hence, it is actually rational. You would have found a similar situation for questions 1 and 2. In fact, infinite repeating decimals *are* rational numbers, as will be shown in the following example.

Mathematicians use various techniques to convert repeating decimals to fractions. These might be interesting enrichment topics for students, although the algebraic methods demonstrated below may be more advanced than the typical elementary curriculum; however, given the prevalence of calculators, which result in repeating decimal representations of fractions turning up frequently, it can be helpful for teachers to know that there *are* ways to convert repeating decimals to fractions. The technique for converting 0.3333… to a fraction is illustrated below.

Example

Let *x* represent the value in question, so $x = 0.33333\ldots$. Because the repeating part of the decimal is one digit long, we start the conversion method by multiplying the value by 10. (For every digit in the repeating part of the decimal, multiply by another 10. For example, if the number is 0.123123123…, we need to multiply by 1,000 instead of 10.)

The general idea of the method is to compare 10 (or whatever number of tens we multiplied by) times the given value with the original repeating decimal number. In this example, we see that 10 times 0.33333 is 3.3333…. Because the decimal parts of both of these numbers go on forever, the parts to the right of the decimal are said to be equal. Ten times the number is 3.3333…. If we subtract one of the 0.33333s… from 10 of them, we have removed the decimal part and are left with 3. This resultant value is equal to 9 times the original number. Numerically,

$$3.3333\ldots - 0.3333\ldots = 3$$

We can calculate $9x = 3$

Therefore, $x = \frac{3}{9}$ or $\frac{1}{3}$

We have found the fraction equivalent of 0.33333…. We can always use a method like this to show that a repeating decimal is rational.

There are times when working with a fraction, rather than a decimal approximation of a repeating decimal, is preferable. While a good calculator generally treats calculating with a value such as 0.33333… with enough decimal places to be equivalent to using 1/3, difficulty arises when students use a rounded value such as 0.33, as though it is exactly *equal* to 1/3. If we multiply a large number, for example 999,999 by each of 0.33 and 1/3, we see that the result is *not* the same. Judgment is required to determine when using a rounded value is good enough.

As students progress in mathematics, they will encounter still other types of numbers. For example, the square roots of negative numbers, such as $\sqrt{(-1)}$, are called *imaginary numbers*. While students do not generally need to use numbers such as these in elementary school, it might be helpful to the teacher to have an answer prepared in case a student asks, "What if you try to take a square root of a negative?" Centuries ago, mathematicians were shocked to find that such numbers existed, and even tried to cover up the existence of some of the stranger mathematical numbers to keep mathematics pure. Today, mathematicians are more comfortable with the idea that some numbers are difficult to model with concrete objects. The mystery and intrigue of mathematics beckons those who are willing to live with such surprises!

> **MATHEMATICAL TERM**
>
> An *imaginary number* is a number that has a square root that is less than zero. For example, $\sqrt{-25}$ is an imaginary number and its square is -25.

Chapter Problems

1. Enter any whole number between 2 and 9 into your calculator. Divide that number by 2. Divide the result by 2 again. What happens if you do this 10 times? 20 times?
2. Enter the same number you used in Problem 1 into your calculator. Divide it by 0.5. Divide the result by 0.5 again. What happens if you do this 10 times? 20 times?
3. Considering what happened when you performed the calculations outlined in Problems 1 and 2, predict what would happen if you repeated the operation 100 more times in each case.

Further Reading

Kajander, A. (2013). How close can we get? Discrete and continuous data. *OAME/ AOEM Gazette, 51*(3), 28–34.

Different types of numbers are described in the context of data

Fischer, C. A. (2013). *Beyond infinity: A matheMATTical adventure.* Littleton, CO: Sienna Books.

This engaging novel is suitable for middle-school students and adults alike, and combines science fiction concepts and real mathematics. See www.kickstarter.com/projects/504814430/beyond-infinity-a-mathematical-adventure for an introduction by the author.

Gadanidis, G. (2012). Why can't I be a mathematician? *For the Learning of Mathematics, 32*(2), 20–26.

This article provides a glimpse of the excitement and intrigue experienced by mathematicians as they work in their field. Engaging classroom activities illustrating complex topics, such as infinity, are provided.

Chapter 11
From Patterns to Algebra

11.1 Designs and Patterns

One of the most powerful aspects of mathematics is its ability to describe quantifiable patterns and make further predictions based on what has happened in the pattern to that point. An essential aspect of a mathematical pattern is predictable repetition or change.

While it is important to look at real-world patterns, as well as to have children examine and create their own geometric patterns, it must be remembered that it is the notion of predictability that gives mathematical patterns their importance. While a picture of a geometric design may be pleasing to the eye, it may or may not contain a pattern that can be replicated beyond that particular picture. On the other hand, if floor tiles are used to create a geometric design, and these tiles are arranged with predictable repetition, then a pattern is created. Examining a real-world pattern to determine what comes next, and to see what occurs in the pattern much further along, focuses our attention on prediction, which is the aspect of patterns that mathematics can best address.

11.2 Exploring Patterns with Contexts and Concrete Materials

In higher-level mathematics, algebraic symbols and expressions are often used to describe patterns and relationships. Before introducing symbolic representations, it is important for students to develop a more visual and concrete sense of different kinds of patterns. Once they can recognize and informally describe patterns, more formal or algebraic descriptions can emerge. Introducing algebraic descriptions of concepts and relationships that children have *not* visualized or understood concretely contributes to algebra's reputation of being mysterious and confusing, and thus incomprehensible.

There are many real-world contexts that can be used to help students study patterns. Walking into a fabric or wallpaper store, or even examining wrapping paper, makes this obvious. These contexts provide inexpensive connections to geometric patterns. Once a selection of patterns is available, it is helpful for students to not only look at them, but to use the visible parts of the patterns to make predictions about how the pattern would continue. The sample problem to follow gives a sense of this idea, but tailoring the problem to a specific concrete example of interest to your students would improve it. Note that the exploration could involve numeracy, geometry, and measurement, as well as patterning. Allowing children to share their thoughts verbally and by using diagrams or samples of real-world objects and meaningful contexts supports important learning about the underpinnings of predictable patterns. This is a central message inherent in this chapter.

Exploration/Task

Imagine you are looking at a 0.5 metre length of wallpaper in which you have measured the repeated part of the pattern to be 20 centimetres, as measured along the length. If you want to join up a new piece to the 0.5 metre strip, where would you have to start in the pattern to seamlessly continue the pattern?

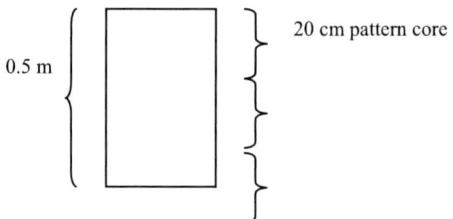

Follow-Up and Discussion

Drawing a model is an important first step in patterning. Looking at this model makes it clear that in 0.5 metres (50 centimetres) there are two complete lengths of 20 centimetres, plus another 10 centimetres. This 10 centimetres is half of one section of the pattern, meaning that the wallpaper stops halfway through the repeated pattern—that is where the next piece must begin. To join the pattern exactly, we would have to start the new piece in the middle of the pattern, measured along the wallpaper length. To answer the question posed in the above task, it is important to know the size of the part of the pattern that is repeating, which is called the *core*.

Unless we are told that the core is repeating, as we were in the previous task, we need to see at least three repetitions of the core to be able to identify it, and to know that the pattern is not alternating, growing, or changing in some other way. Examining concrete examples of simple repeating patterns is an excellent way to help students begin to develop the skills to understand more sophisticated patterning and algebraic concepts.

> **MATHEMATICAL TERM**
>
> The *core* of a repeating pattern is the part or attribute that is repeating.

11.3 Connections between Geometry, Patterns, and Algebra

The topic of patterning is typically grouped with algebra in the curriculum, preceding more formal algebraic study. But because geometry often plays a key role in patterning, we recommend threading the ideas of patterning through other strands. Focusing on the mathematical aspect of these patterns—the idea of prediction—allows students to make mathematical sense of the concrete examples, and supports the subsequent development of more sophisticated algebraic concepts.

Follow-Up and Practice

Explore the following patterns. Construct a representation (numeric or geometric) to help you understand how the pattern is changing. Use your representation to help you answer the questions.

1. Assume that the pattern of numbers below continues in the same manner. What would be the next three numbers in the pattern?

   2   5   8   11   14   17   20   23

2. A necklace is made with tubular beads that are 2 centimetres long. The following pattern is repeated: green, silver, brown, gold. If an equal number of each bead is used, and no cord is showing, what are the next three possible necklace lengths longer than 40 centimetres?
3. A fence consists of a 10-centimetre-wide post, then a 30-centimetre-wide piece of lattice, then a 10-centimetre-wide post, then a 30-centimetre-wide lattice, and continues this way. If the fence must end with a post, what is the minimum possible fence width that is closest to, but more than, 2 metres, assuming that the widths of the posts and the lattice are not altered?

11.4 Patterns and Number Properties

Due to the inherent patterns in the number system, a deep connection between numeracy and patterning is possible. Focusing on this connection can enhance students' understanding of both ideas. A simple starting point is the place value system for decimal numbers. We have seen that as we move to the left in a decimal number, the digits represent a quantity 10 times greater than the number to the right.

33,333
↓
Each 3 represents a value 10 times greater than the one to its right.

156 Chapter 11

> **CONNECTION**
> To review the use of base ten blocks with whole numbers, see section 3.3.

In each location (column) in a number, the same pattern is used to increase values. Numbers from 0 to 9 are used, followed by starting again at 0 and adding 1 to the larger column, and so on. A hundreds chart is a valuable model for students when they are trying to see such patterns.

A pattern that is observed less often is the geometric repeating pattern of the base ten blocks when used to show increasing values. For example, 1, 10, and 100 can be represented as:

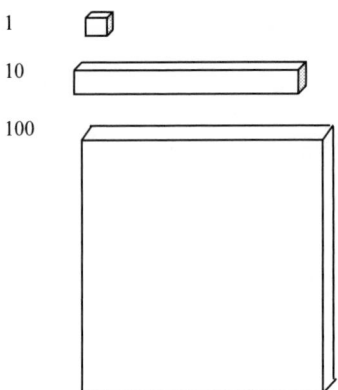

Using this system, 1,000 can be represented by a cube ten times as great in *all three* dimensions as the centicube (or the 1). Another way to think about the thousands cube is that it is made up of ten layers of hundreds pieces. If we are going to continue to represent larger and larger values by sketching representations of 10,000, 100,000, and 1,000,000 using drawings of base ten blocks, what would they look like? (You might want to use a chalkboard for this.) For example, if we use the 10 × 10 × 10 centimetre cube to represent 1,000, then *ten* of these cubes would be needed to represent 10,000. (Note that if these thousands cubes are lined up in a long row, they form a shape that resembles a larger version of the long or rod.) How could you represent 100,000? How do these three new shapes (to represent 1,000, 10,000, and 100,000) compare to the first three shapes (which represent 1, 10, and 100)? Is there a pattern? Can you visualize what shapes might be created if you were to use the base ten blocks to represent 1,000,000, 10,000,000, and 100,000,000? Such visual thinking might help some students see how much larger the numbers get, and recognize the inherent pattern in the decimal system and these models of it.

Patterns can also help students to understand operations, particularly multiplication. As was discussed in the context of whole number multiplication, patterns can help students with the times tables, too. They can be encouraged to look for patterns in the times tables chart, which can strengthen both their knowledge of facts and their pattern recognition skills. Asking students to examine copies of the times tables chart to see how many number patterns they can find is an interesting activity.

> **CONNECTION**
> To review whole number multiplication facts, and examine patterns in these facts, see section 5.8.

11.5 Repeating Patterns and Pattern Core

Consider the following arrangement of shaded squares:

How would you repeat this pattern? Is it a pattern? Does this arrangement provide enough information for you to know for certain?

Exploration/Task

Using the above arrangement of squares, draw more than one pattern that starts with the same four squares shown here. Extend each pattern far enough that the pattern is clear.

Follow-Up and Discussion

To complete the above task, you might have drawn

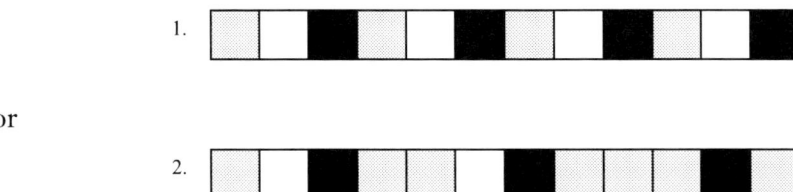

or

Which pattern is correct? There was not enough information provided in the initial diagram to know for certain. Both of the above patterns begin with the same four squares, as was specified in the task, and both show a repeating pattern. In pattern 1, the repeating part or core has three elements, namely grey, white, and black, and the next grey square is assumed to be part of the next repeated section. In pattern 2, the core has four elements, namely grey, white, black, grey, and this (four-square) pattern block is repeated. When analyzing repeating patterns, enough information is needed to be reasonably certain of the pattern. At the very least, two full repetitions, but ideally three, of the complete core are needed, and we need to know—or we must conjecture—that the pattern continues in the same manner.

11.6 Growing and Shrinking Patterns

Of course, in the real world, as well as in mathematics, not all patterns repeat over and over again in exactly same way. A pattern can, for example, grow or shrink, unlike patterns that repeat the same core over and over (as we would expect of wallpaper, for example). Consider the following:

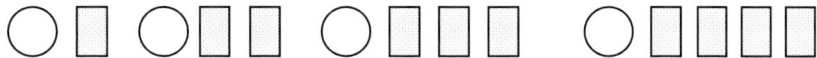

You might predict that the next element in this pattern will have five rectangles next to the circle, because you have noticed that the number of rectangles increases by one each time. (Note that the phrase "with each iteration" is sometimes used, rather than "each time.")

In contrast to this example, the following number pattern is shrinking, rather than growing: 100 95 90 85 80 75. One way to describe this number pattern might be: "start with 100 and subtract 5 each time." This description allows us to find the next term fairly easily, but is not very helpful if we want to efficiently find a number quite a bit further along in the pattern. Mathematics can help us find more efficient ways of making such predictions.

11.7 Making Predictions about Linear Patterns

In this section, we will use some simple contexts to develop ideas about linear patterns.

Example

Your school has square tables available for events. Normally, 4 chairs fit around a table:

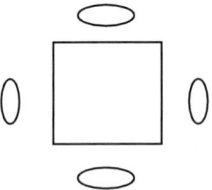

For the year-end dinner, tables will be lined up in a long row in the gym. For every 2 tables, there will be a total of 6 chairs. Here is the chair arrangement for 3 tables:

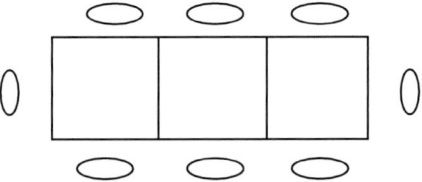

One way to keep track of this information is to use a table that compares tables to chairs:

Tables	Chairs
1	4
2	6
3	8
4	
5	

Exploration/Task

Fill in the number of chairs in the chart for 4 and 5 tables, then continue with the questions below.

1. Think about how you knew how to determine the number of chairs that would be used for 4 and 5 tables. Continue the pattern for 6 and 7 tables. Do you still need to draw the tables and chairs to be able to fill in the next line in the chart?
2. Without filling in more values in your chart, find a way to calculate how many chairs fit around 10 tables arranged in the same way. Find more than one way to calculate this number of chairs.
3. The largest number of tables that fit in a row in the gym is 15. How many chairs can be arranged around these 15 tables? Again, try to solve the problem without referring to the chart.
4. If 28 people are coming to the dinner, how many tables are needed? Solve this problem in more than one way. If possible, share and discuss your methods with a partner.

Follow-Up and Discussion

You may be surprised at what you find when you share your thinking and solutions with a partner. The richness of this problem is in the various possible solutions, and how they relate to each other. These possibilities are why classroom discussion and sharing after an activity is such a crucial aspect of conceptual learning.

Problem 1 could have been completed by adding 2 more chairs for each new table. A number of different types of reasoning might have been used to solve Problem 2. If you do not have colleagues with whom to share, consider the following possible student solutions for Problem 2. Work to follow the reasoning in each case, and compare it with your own reasoning.

- I knew from the smaller number of tables that 2 chairs are added for every new table. I already knew from my chart that 7 tables had 16 chairs, so I added 2 chairs for each of the 3 new tables after 7. The number of chairs for 10 tables is 16 + 2 + 2 + 2 or 22.
- I looked for a relationship between the number of tables and the number of chairs. Leaving off the chairs on each end for now, the number of chairs is 2 times the number of tables. I then added 2 for the chairs on each end. The answer is 10 × 2 + 2 = 22.
- I pretended that each of the 10 tables had 4 chairs. That's 10 × 4 = 40. But there aren't any chairs between the tables (where they join). So for every spot that they join, we lose 2 chairs. For 10 tables, there are 9 joins (I almost thought it was 10, but when I drew it I saw it was 9). I subtracted 9 joins × 2 chairs or 18 chairs. The answer is 40 − 18 = 22.
- One table has 4 chairs. When I add a new table, I add 3 chairs, but I lose the one where the tables touch. For 2 tables, it is 4 + 3 − 1. For 3 tables, it is 4 + 3 − 1 + 3 − 1. For 4 tables, it is 4 + 3 − 1 + 3 − 1 + 3 − 1. I do the "+ 3 − 1" calculation for each of the 9 new tables. The answer is 4 + 9 × (3 − 1) = 4 + 9 × 2 = 4 + 18 = 22.

The methods that are similar to your own will be easier for you to understand than those that use different reasoning. Similarly, others may have to work to understand your thinking! The art of being an effective teacher is developing the skill of understanding *others'* solution methods, rather than just the one that you prefer. A method that is easy for one person to use and understand is not always the easiest for another. It is challenging to move away from our own preferred solution method and embrace another, and to realize that another person's chosen method might be the easiest way for *them* to think about the problem.

Practice using a solution method that is not your own. Choose one of the solution methods provided in solutions 1 to 4 that you did not think of yourself, and apply this reasoning to Problem 3 of the task. Of course, you should get the same answer as you did using your original method! Compare the thinking and reasoning involved.

For Problem 4, a number of possible methods can be used to solve the problem. For example, a chart could be used, which would be read *backwards* from 28 people to find the number of tables. If you already know the number of people who could be seated at 15 tables, you might work backwards from that number because the two answers will be relatively close. Since these numbers are relatively small, a number of trial-and-error strategies, or working backwards strategies, work quite well. Alternately, you might simply have started with 28, subtracted 2 from the ends, and divided the rest by 2, to get 13 tables. When patterning, making note of particular strategies and working to expand our available strategies are important activities. If the values become very large, however, more sophisticated methods can save time. The remaining sections in this chapter will develop some of these more formal methods and strategies, beginning with the method of graphing.

11.8 Graphing Linear Patterns

Students who have played games such as Battleship know that two numbers can be used to determine a location on a flat surface. They just need to determine which number measures the distance in which direction. In mathematics, the following order is standard (although the other way around will also work, as long as this change is made clear).

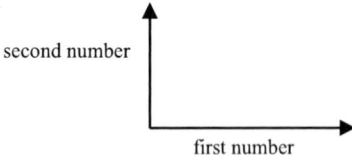

It is also standard to start from (0, 0), which is located at the bottom left corner of a grid or graph if we are using whole (or positive) numbers. If there are also numbers less than zero, we add to the grid to the left and beneath the (0, 0) location. For example, the location (3, 4) on a grid is taken to mean 3 to the right and 4 up:

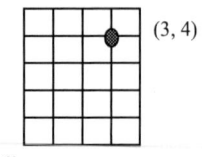

Grids and graphs can be used to record patterns because there are often two numbers involved in a pattern—one that tells us where we are in the pattern and one that represents the answer or result. Because grids and graphs allow us to see the relationship between two sets of values, they are a useful way to record patterns and can provide more information than number charts in terms of what we can expect to happen next. For example, the situation outlined above that asked how many chairs were needed for a specific number of tables can be represented graphically. We need some way of numerically keeping track of where we are in the pattern. For simple geometric patterns, we might number the pictures, starting with 1, 2, 3..., in which case we can determine where we are in the pattern using this picture number. A numeric pattern might similarly be organized by term number—a sequential ordering showing each result—in a list. For the table and chairs example, the term *number* might refer to the number of tables, which is in turn used to determine the number of chairs. The phrase *input number* is sometimes used to refer to the term number or picture number, especially if the pattern is organized in a chart or with a pattern rule. The picture number, input number, or term number is traditionally counted on the horizontal axis. The result in each case is recorded on the vertical axis. The two values together determine the placement of each point.

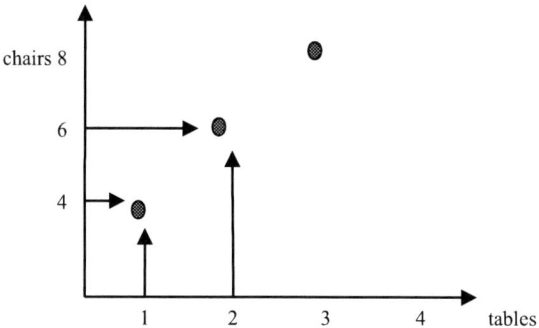

The same observations can be made about the situation from the graph as from the table provided above, but a graph often makes trends and relationships more obvious. For example, we don't need to do any calculating to see from the graph that the number of chairs increases by the same amount for each new table (i.e., by 2). It is also clear that this increase will continue with each table added. At some point, though, we will run out of space to plot very large numbers on the graph. According to the information provided (see section 11.7), the greatest number of tables that will fit in a row in the gym is 15. In this situation, the number on the horizontal axis of the graph (which represents the number of tables) cannot be greater than 15. As described in section 10.3, a word used in mathematics to describe such a restriction is *domain*; in this case, we would say that the *domain* of the pattern is whole numbers up to 15. (If the situation has no constraint or end value, we do not need to specify a domain, or we simply say the pattern holds for all whole numbers or whatever numbers are relevant for the problem.) Similarly, the *range* is used to describe the set of resultant values in the pattern; in this problem, the range represents the possible number of chairs for the situation (up to 32). When these words are used unnecessarily in number patterns, they can cause students a lot of confusion. We recommend continuing to provide realistic situations as a way to help students make sense of the need to (sometimes) specify the domain and range.

11.9 Pattern Rules

In this section, we will explore another example of a geometric growing pattern using a chart and a graph, and will begin to discuss how pattern rules can be constructed and understood, and what they might tell us.

Exploration/Task

Consider the following pattern made from toothpicks.

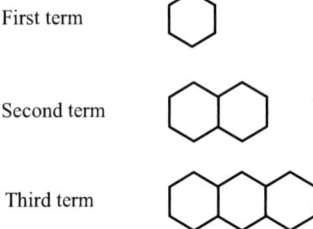

Next to each shape or term, write down the total number of toothpicks needed to build the shape, then summarize this information in a chart. Charts like this are sometimes called input/output charts. The *input* number is the picture number or term number (which describes where we are in the sequence), while the *output* number is the value of whatever we are counting—in this case, the number of toothpicks.

Follow-Up and Discussion

Creating the next shape (or term) in the pattern model allows us to count the number of toothpicks needed for each; however, at some point, this becomes tedious. When you completed the task, you may have observed that the number of toothpicks in each term increases in a pattern: the output value increases by 5 each time. This corresponds to the additional number of toothpicks required to make the next shape.

Input	Output
1	6
2	11
3	16
4	21
5	?

One of mathematics' strengths is that it allows us to keep track of predictable patterns and relationships in a concise way. In addition to using a table, we could also use a graph to keep track of how the toothpick pattern is growing. Reading the graph allows us to predict the number of toothpicks further along in the pattern.

Exploration/Task

Create a graph for the first three pictures in the toothpick pattern shown above. Record the input number, picture number, or term number on the horizontal axis, and the output number (number of toothpicks) on the vertical axis. Use your graph to predict the number of toothpicks in the fourth term (or picture four). If you knew the total number of toothpicks in a given shape, how could you use your graph to determine the picture number?

Follow-Up and Discussion

Your graph should have the points (1, 6), (2, 11), and (3, 16), with the first number used to show the distance on the horizontal axis. Notice that each new point is 5 units higher than the previous, which tells us that the next output value must be 21. Alternately, if we know the number of toothpicks (which is the number on the vertical axis), we can read across and down to find the corresponding input number. We have observed that we can find the next output value in a pattern by looking at a chart or a graph and noticing how much the pattern changes. Can we determine an output number without going through all of the preceding numbers in the chart in order?

Exploration/Task

Using the toothpick example, we see that there seems to be a relationship between the number of hexagons and the number of toothpicks needed. If we are given the number of hexagons, for example 10, how can we count the number of toothpicks needed? Look for a relationship between the number of hexagons and the required number of toothpicks, and determine a rule that will allow you to calculate the number of toothpicks needed to create a certain number of hexagons. Test your rule on two different pairs of values that you know from the input/output chart.

Follow-Up and Discussion

You might have found a number of ways to determine a rule, depending on how you visualized the pattern construction. Here are a few examples:

1. If you visualized the pattern as *5 toothpicks added for each hexagon*, you might have seen the pattern like this, which uses the third term:

    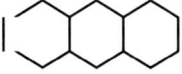

    The toothpick furthest to the left is isolated, as it is a special case needed for the first hexagon only. We need to add it on just once. As a result, the rule is: *Start with 1 toothpick, then add 5 more for each hexagon*. We could write the rule more formally as: 1 + number of hexagons × 5.

2. Another way of seeing the hexagons is to think of them first as separate hexagons. Using this idea, we might see the third term as

That gives us 6 toothpicks for every hexagon; however, when we slide the shapes together, we realize that the overlapping toothpicks have to be removed. The number of overlaps is one *less* than the number of hexagons. In this case, the rule is: *6 toothpicks per hexagon, minus the overlaps*, or 6 × (number of hexagons) − (number of hexagons −1).

Many other possible rules could be used for this example. It is a great challenge for teachers to get inside students' heads to understand the reasoning they used to create their own rules. Supporting students' own reasoning when constructing such rules is much better than showing students a rule that may not connect to how they visualized the pattern. We will continue to formalize these rules in the next section.

Practice and Further Exploration

1. Instead of hexagons, use joined squares to make a similar pattern with toothpicks. Here are the first three pictures.

   Find at least two rules to connect the number of squares to the number of toothpicks. Use the fourth picture to test your rules.
2. For the hexagon pattern explored in this section, you came up with two new rules that allowed you to determine the number of toothpicks needed to create the pattern. Draw the third hexagon picture determined by each rule, and connect each part of the rule to a geometric part of the picture.
   a) (4 × number of hexagons) + (number of hexagons + 1)
   b) 6 + 5 × (number of hexagons −1)

11.10 The Variable

Early grades story problems may result in mathematical equations such as $5 + ? = 8$ or $5 + x = 8$. In this case, the ? or x is called an *unknown*; here it is a placeholder for one unknown number, and only *one* numeric solution makes a correct equation.

In the example presented in section 11.7, a pattern was created by counting the number of chairs that could be placed around different numbers of tables. We saw that there was a relationship between the number of chairs and the number of tables. One aspect of the problem was that there was uncertainty about the number of tables, and, hence, about the number of chairs; as a result, we had to examine the problem for a *varying* number of tables and chairs. For problems in which there is such uncertainty, we often refer to the unknown numbers as *variables* because there are many possibilities for their values; in other words, they can vary in value. This change in the use of a letter such as x to represent *many* different possible values (rather than a single value) can be a conceptual roadblock for some students.

The following is an example of a case in which the symbols x and y are used to represent number of possible values: $x + y = 10$. Each variable can take on a number of values, but they also depend on each other. It is important to keep in mind that sometimes a letter such as x is used to represent only one unknown value, but at other times it can represent a range of values.

> **MATHEMATICAL TERM**
>
> A *variable* is a symbol used to represent a value that can vary.

> **INSIGHT**
>
> A variable may have many possible values, depending on the context.

Practice and Further Exploration

Assuming that x and y are positive whole numbers, find all the (x, y) pairs that solve $x + y = 5$. Graph these solutions as points on a grid. What do you notice?

11.11 Equation Concept

In the tables and chairs example explored earlier in the chapter, we saw that a relationship existed between the number of tables and the number of chairs. There is a chair on 2 sides of *each* table (so the number of these chairs was 2 times the number of tables), and 1 chair on each of the 2 ends of the row of tables (added just once). The following model illustrates this context:

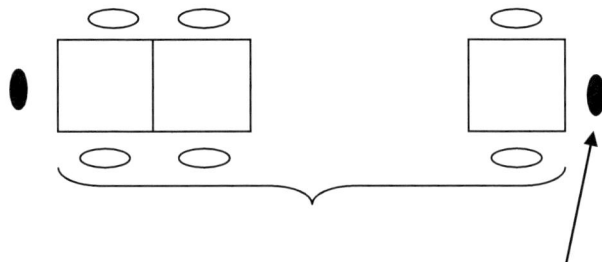

twice the number of chairs as tables *plus two* extra chairs (shown in black on each end)

A general model of the problem is useful when attempting to find the relationship or mathematical rule. The conceptual difficulty for students is this very step—the connection between the idea of 3 tables, 4 tables, 5 tables or *any* number of tables, where the *any* number is represented by x or n. Students who are not conceptually ready to think generally about the relationship will still tend to want to find a single value that "works" for x. If students are not ready to see that x can take on a number of values, they need to work through more examples, and examine more diagrams like the one above before proceeding. It can also be helpful to explore the pattern as the number of tables increases:

Number of tables	Calculation for the number of chairs
1	$2 \times 1 + 2$
2	$2 \times 2 + 2$
3	$2 \times 3 + 2$
4	$2 \times 4 + 2$
10	$2 \times 10 + 2$
n	?

Exploration/Task

Use a letter such as n to represent the number of tables in the example, and use the information provided to find a mathematical rule to calculate the number of chairs needed for *n* tables. Test your rule with 1, 2, and 3 tables to be sure it calculates the correct number of chairs in each case. If you can, find a second rule that will also allow you to calculate the number of chairs for a given number of tables. Refer back to the informal thinking outlined in the sample solutions in section 11.7 if necessary.

Follow-Up and Discussion

Using a model, you may have been able to see that for any number of tables *n*, there are *twice* as many chairs. Therefore, $2 \times n$ calculates the number of chairs along the sides of the tables. We then add one chair on *each* end, which is 2 more. These two ideas allow us to generate the rule for calculating the number of chairs: $2 \times n + 2$. (Of course, there are other ways to write this rule, as we saw with the hexagon pattern in section 11.9.) If we use another letter, for example C, for the number of chairs, the rule becomes $C = 2 \times n + 2$. This rule represents a *relationship* between C and *n*: each *depends* on the other, and both can vary.

In section 11.8, we saw that the graph of possible values for C and *n* formed a straight line. For this reason, it is called a *linear relation*. The mathematical rule allows us to find the value of one variable (C or *n*) from the other. If one value changes, the other will also change.

> **MATHEMATICAL TERM**
>
> A *dependent variable* depends on another variable to take its value. For example, in the relationship $y = 2 + x$, the value of *y* depends on the value used for *x*. Here, *x* is the *independent variable*. It represents the location in the pattern. The term *independent variable* is a more formal name for the term number, and may not be a whole number.

Practice and Further Exploration

1. Your school has a few gym lockers available, but not enough for everyone. The student council decides to rent the lockers for a fee to those students who want one, and to use the money to pay for school trips. They have some extra locks, so they decide to require students who wish to use a locker to purchase one of these locks for $8. They set the monthly rental rate at $3. Write an expression to calculate the total cost, C, of a locker for *n* months.
2. Sarah's mother gives her $20 to put towards a gym locker for the year. Will this cover the cost of the locker for the entire (10-month) school year? If not, how many months will it pay for? Solve this problem in several ways—for example, you could use a chart or a graph.

11.12 Solving Equations with Manipulatives

When solving Problem 2 in the previous section, you might have set up an equation like this: $C = 8 + 3 \times n$. You could have used a chart or a graph, or even trial and error, to find that Sarah's mother's $20 contribution will pay for the first 4 months of the locker rental. We can double-check this by calculating the lock cost of $8, plus the locker rental for 4 months at a cost of $3 each ($12), which is $20 in total. The statement $C = 8 + 3 \times n$ shows a relationship (between the two expressions C and $8 + 3 \times n$) and states that the expressions must be *equal*. The term for this type of statement in mathematics is *equation*.

> **MATHEMATICAL TERM**
>
> An *equation* is a statement of equality between two expressions. It contains an equal sign.

Exploration/Task

Continuing with the locker example, if we know n (in this case the number of months), we can calculate the total cost. Calculate the total cost of renting the locker for 1 year (10 months).

Follow-Up and Discussion

You might have quickly multiplied 10 times $3 and then added $8. A more formal written method in mathematics involves showing the substituted number in the expression using brackets:

$$\begin{aligned} C &= 8 + 3 \times n \\ &= 8 + 3 \times (10) \\ &= 8 + 30 \\ &= 38 \end{aligned}$$

The total cost for 10 months is $38.

What if we want to solve the problem the other way around? That is, what if we know how much money we have, but not the number of months it will pay for? Following up with Problem 2 from the previous section, we had $20 available and wanted to know how long that would last. We know C, but not n. In formal notation, the problem becomes $20 = 8 + 3 \times n$.

We have discussed several possible methods that can be used to solve such equations, including a table of values, a graph, or even trial and error. At some point, if the numbers we are dealing with get really big, these methods will be cumbersome. In the next section, we will investigate other ways of solving equations like this one.

Models for Solving Equations

The equal sign is a mathematical symbol that tells us that the two sides of the equation must (at all times) be equal or balanced. We can use the metaphor of a balance, or even model the equation using a real pan balance and weights such as centicubes, to allow students to experience the meaning of the equal sign. (Too often students simply equate the equal sign, =, to the answer button on a calculator, rather than understanding the required balance of everything on either side of it.)

As discussed, the n represents an unknown amount that can vary. In this case, since we now know what C is (if C is $20), there will be only one possible correct number of months (n) for this cost value. The symbol n now represents a mystery amount that makes the equation true in this case. To model this situation concretely, we can think of $3 \times n$ as three of the same mystery amounts or bundles that we are not allowed to open. We could find the value (or the number of centicubes) in each mystery bundle using the balance idea. Here is a visual image, in which each little cube represents one unit, or here, $1.

It is tempting to use trial and error to test the number of cubes in each bundle until we get a true statement. But once we have set up an *equation* (a relationship with an equal sign), we only have this true statement if both sides remain the same (i.e., equal) at all times. We must follow the basic principle or rule that says that, whatever we do, both sides have to remain equal at all times. Following this equality principle allows us to develop a method that will also be helpful when solving more difficult equations.

Exploration/Task

Using the diagram above, or the same situation modelled with actual manipulatives, follow the equality principle (both sides have to stay equal) and solve the equation concretely. (If you already know about solving equations using algebra, think only about how you would concretely model each step of the solution.)

Follow-Up and Discussion

Two steps are required to solve for the unknown bundle amount. First, we need to remove 8 units from each side.

This results in 12 units remaining on one side, and 3 mystery amounts on the other side. At this stage, the model still illustrates 3 unknown bundles (in the algebraic expression, this is represented by $3 \times n$). The next step is to find the weight or value of 1 bundle. It is tempting to want to subtract bundles, but until we know their weight we don't know how many corresponding cubes to remove, and would violate the equality principle. The key idea is to focus on what we know about operations—if we know that 3 equal bundles add up to 12, how do we find the weight or value of each bundle? The idea of grouping the 12 units into 3 equal groups leads us to the concept (and operation) of *division*. Dividing both sides by 3 in the model results in the following:

While each side now contains one-third of what it did previously, the two sides are still equal and the remaining unknown value is unchanged. We see that 4 units remain on one side and only 1 bundle on the other; therefore, the bundle must have a value of 4. In our context, this means that $20 will pay for a 4-month rental.

We can't stress enough the importance of developing equation-solving methods based on concrete reasoning and understanding. The arbitrary imposition of (sometimes incorrect) rules such as "add the opposite amount to the other side to 'cancel' a value" is to be avoided at all costs as, later on, such rules may not generalize and can cause much confusion. The development of conceptual understanding is imperative in order for students to continue to construct new knowledge and develop new techniques based on those they already know. Instead of rules, students can develop concrete reasoning for solving equations that will indeed generalize.

If you already know the algebraic steps for solving an equation like $20 = 8 + 3 \times n$, you might have been able to make a connection between the action in the model and the (underlined) algebraic steps to follow. If you are not familiar with the algebraic method, note that it will be further explored in the next chapter.

$20 = 8 + 3 \times n$

$20 \underline{- 8} = 8 \underline{- 8} + 3 \times n$, which, since $8 - 8 = 0$, simplifies to

$12 = 3 \times n$

Next, we group each side into three equal groups, using division:

$12 \underline{\div 3} = 3 \times n \underline{\div 3}$

Recall that we can multiply in any order, so we can rearrange the right side:

$12 \div 3 = 3 \div 3 \times n$, and since $3 \div 3 = 1$,

$4 = n$

The algebraic method is only mentioned here to allow readers already familiar with it to make the connection between the formal method and the concrete one. However, it is best to teach the formal method later when working with students. As always, it is best to allow the more formal technique to evolve as a shortcut for the concrete method when students are ready.

Practice and Further Exploration

Using a drawn model that shows equality (keeping in mind the idea of a balance), model and solve $2 \times n + 4 = 14$ for n. Be sure that your model preserves the equality of both sides of the balance or expression at all times. If you are already familiar with the algebraic method, think about how your actions align with the steps.

Chapter Problems

1. Students are fundraising for cancer by participating in a run. Their sponsors have agreed to pay them an initial $5, plus $2 for each kilometre they run. Use a chart, diagram, or graph to determine how much money sponsors will contribute to the fundraising campaign if the students run:

    a) 5 kilometres
    b) 10 kilometres
    c) 20 kilometres

2. Write an equation that includes a variable that could be used to determine the amount of money the students' sponsors will contribute for each distance specified in Problem 1. In your equation, what does the variable represent?

3. Compare the answers you found for students running 20 kilometres using the chart or graph in Problem 1 and the equation in Problem 2—your answers should agree. If students ran 100 kilometres, which method would it be easiest to use to determine the total funds raised?

Further Reading

Beigie, D. (2011). The leap from patterns to formulas. *Mathematics Teaching in the Middle School, 16*(6), 328–335.

    Beigie uses three-dimensional shapes to explore patterns and create algebraic formulas.

Ferrini-Mundy, J., Lappan, G., and Phillips, E. (1997). Experiences with patterning. *Teaching Children Mathematics, 3*, 282–288.

    Using the pattern of a swimming pool with a border, the authors illustrate how algebraic concepts, including reasoning, build from Kindergarten through Grade 6.

Hawes, Z., Moss, J., Finch, H., and Katz, J. (2012/2013). Choreographing patterns and functions. *Teaching Children Mathematics, 19*(5), 302–309.

    Hawes, Moss, Finch, and Katz build concepts of algebra and algebraic rules with early elementary students based on the patterns inherent in dance. The article discusses creating picture patterns and input/output chart activities, and builds toward students creating dance patterns of the functions.

Kaplan, R. G., and Alon, S. (2013). Using technology to teach equivalence. *Teaching Children Mathematics, 19*(6), 382–389.

    Kaplan and Alon share professional development targeted at elementary teachers as they use technology to promote understanding of the equal sign in equations. The article discusses the use of the pan balance model, both concretely and virtually.

Markworth, K. A. (2012). Growing patterns: Seeing beyond counting. *Teaching Children Mathematics, 19*(4), 254–262.

    Markworth uses geometric patterns to develop algebraic concepts and discusses how to support students in this development.

Chapter 12
Algebraic Concepts

12.1 Solving Equations Algebraically

CONNECTION

This section formalizes the concepts presented in section 11.12: Solving Equations with Manipulatives, which should be read before continuing.

In section 11.12, a balance was used to concretely model solving equations. We explored how manipulatives can be used to model equations containing positive numbers, such as $2 \times n + 4 = 14$. The algebraic steps are simply a written version of the physical actions performed when using the concrete objects. If students have explored the logical physical moves required to isolate an unknown amount on one side of a balance (or equation), there is no need to teach them formal rules, as these become obvious. Developing concepts with understanding ensures that students will be able to apply these concepts to more difficult equations.

Exploration/Task

Review your understanding of the concrete actions used to solve an equation by drawing a picture that represents each step below. Review section 11.12 as necessary.

$$2 \times n + 4 = 14$$
$$2 \times n + 4 \underline{- 4} = 14 \underline{- 4}$$
$$2 \times n = 10$$
$$2 \times n \underline{\div 2} = 10 \underline{\div 2}$$
$$n = 5$$

Follow-Up and Discussion

In your model, a first step would be to show 2 unknowns and 4 units on one side, and 14 units on the other. We begin by removing 4 units from each side, leaving 2 unknowns on one side and 10 units on the other. Dividing by 2 (grouping each side into 2 equal amounts and taking one amount of each) results in 1 unknown on one side, and 5 units on the other.

The key idea in solving equations—a concept that generalizes and can therefore be used to solve much more complicated operations—is that the *same* operation must be performed on both sides. This preserves the equality of the sides and does not change the value of the variable. Although this results in a different equation, the relationship between the variable and the numbers stays the same. For example,

> **INSIGHT**
>
> If two expressions are equal, then any operation done simultaneously to *both* of the expressions will result in two (changed) expressions that are still equal to each other. The value of the unknown or variable is unchanged.

$2 \times n + 1 = 7$ and the equation we get after subtracting 1 from each side, $2 \times n = 6$, look different, but, in each case, we can see that n must be 3. If two things are equal, it makes sense that half of each amount will also be equal, and so on.

The *same to both sides* method of solving equations is important because it also generalizes to much more difficult equations. For example, if we want to simplify an equation containing -17 on one side, we can *add* 17 to both sides of the equation. This will result in $-17 + 17$ (which is zero) on one side, although it may or may not simplify the other side. In general, using the *inverse* operation will reverse or undo an operation.

Example

$$4 \times n - 17 = 3$$

We are trying to invert the operations surrounding n, so we choose to add 17 to each side. Because we know $-17 + 17 = 0$

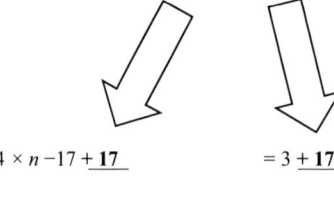

$$4 \times n - 17 + \underline{17} \quad\quad = 3 + \underline{17}$$

$$4 \times n + 0 \quad\quad = 20$$

12.2 Ratio and Proportion

> **MATHEMATICAL TERM**
>
> A *ratio* is a comparison between two quantities of the same units. It can be expressed as a comparison (3:4) or a fraction ($\frac{3}{4}$).

There are particular forms of equations in which each side involves a quotient—in other words, each side of the equation looks like a fraction. Here is an example of the type of equation that is often referred to as a *ratio* problem:

$$\frac{13}{5} = \frac{x}{10}$$

One simple method of solving this problem (to find the value of the unknown x) is to use the idea of undoing each operation as demonstrated above. The inverse operation of the division by 10 on the right side is multiplication by 10, and we need to do this to both sides of the equation to preserve equality:

$$\frac{13}{5} \times 10 = \frac{x}{10} \times 10$$

This simplifies to x on the right side. The left side becomes $\frac{130}{5}$ or 26. (You might instead have calculated $13 \times \frac{10}{5}$, which is 13×2 or 26).

Students sometimes find ratio and proportion a difficult topic. This may be because they have not had the opportunity to understand fractions and equations conceptually. Both of these concepts are related to ratio, and without a solid understanding of them, ratio may be difficult. Students are also sometimes told to use a procedure referred to as *cross-multiplying* to solve ratio problems. As can be seen in the example above, there are one-step methods for solving ratio problems that use only the ideas of solving equations developed in sections 11.11, 11.12, and 12.1.

> **CONNECTION**
>
> To review the related concepts of fractions and equations, see chapters 6 and 7, and sections 11.11 and 11.12.

Exploration/Task

Ben and his older brother Jari agree to put their money together to purchase a new, good-quality bicycle. Jari will use it 5 days a week to ride to the college he attends a few kilometres from home, and Ben will get exclusive use of it on Saturday and Sunday. They agree to share the cost according to the ratio 5 to 2, based on the number of days each will use it.

1. Draw a diagram (a model) to show how they would share the cost if the bike cost $700.
2. Use the diagram to show what *fraction* of the total cost each would pay, and how much that is in dollars.
3. If they sell the bike for $420 at the end of the year, how much money should each of them get? Draw a new diagram showing the $420, and use it to find the amount each would get from the sale. Double-check that the amounts are still in the *ratio* of 5 to 2. Check that your amounts agree with the idea that Ben gets $\frac{2}{7}$ of $420 and Jari gets $\frac{5}{7}$ of $420.

Follow-Up and Discussion

While working through this problem, you may have noticed that a central difference between a ratio and a fraction is how the number of parts in one whole is understood. For example, Ben owns $\frac{2}{7}$ of the bike and Jari owns $\frac{5}{7}$. But the *ratio* of what they own compared to *each other* (rather than compared to the whole) is 2:5. Your model may have looked something like this:

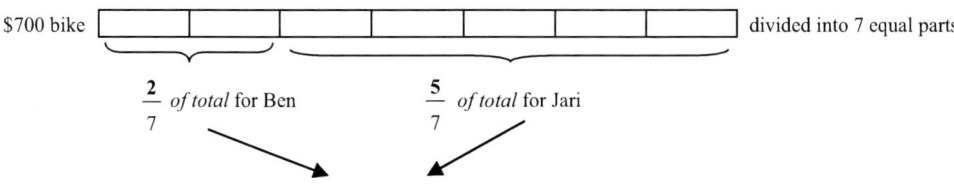

Ratio of amounts to *each other* is **2:5**

How are students to know that the diagram initially needed to be divided into 7 equal parts? This idea requires the concept of a ratio to be understood conceptually; if a ratio is 2:5, then the *whole* must be divided into 2 *plus* 5 parts, or 7 parts in total. Diagrams are key to developing the concept of ratio, and helping students understand how they relate to fractions.

When the $420 is divided into 7 equal parts, we see that each part is $60. Note that the question uses a *multiple* of 7, namely $420, as the selling price, which will result in a whole number when we divide it into 7 parts. It is important for teachers to consider this relationship—and to understand how to generate problems that will result in fairly simple numbers—when constructing classroom questions. Ben gets 2 parts of size $60 ($120 in total) and Jari gets 5 parts of size $60 ($300 in total). We see that $120 and $300 add up to $420 as required. We also see that, using the fraction multiplication procedure (see section 7.3), $\frac{2}{7}$ of $420 is $120, and $\frac{5}{7}$ of $420 is $300. For example, Ben's share is

$$\frac{2}{7} \times 420 = \frac{2}{7} \times \frac{420}{1} = \frac{2 \times 420}{7 \times 1} = \frac{840}{7} = 120$$

Since we can multiply or divide in any order (see section 5.7), we could also have more easily calculated

$$2 \times \tfrac{420}{7} = 2 \times (\tfrac{420}{7}) = 2 \times 60 = 120$$

Sometimes ratio problems are already written as equations. We recommend that starting with a problem and a context is more useful to students, in terms of helping them to understand and construct models.

Exploration/Task

Looking at the bicycle-sharing problem again. What if the problem had stated:

> Ben is paying for $\tfrac{2}{7}$ of a bike that costs $700. How much does he pay?

or

> Ben and his brother share a purchase in the ratio of 2:5. How much will Ben have to pay if his brother is paying $500?

Solve each scenario, while thinking about how they are similar and how they differ.

Follow-Up and Discussion

For the first problem, to determine $\tfrac{2}{7}$ of the cost of a bike that costs $700 in total, you could use a fraction operation such as $\tfrac{2}{7} \times 700$. To solve the second problem, we could use an equation such as $\tfrac{2}{5} = \tfrac{x}{500}$. Although they look quite different, each expression solves the same scenario, but uses different pieces of available information. Each yields the same answer—that Ben's share of the cost is $200.

The equation $\tfrac{2}{5} = \tfrac{x}{500}$ can be solved in several different ways. Using the equation-solving concept explored in sections 11.12 and 12.1, we would reverse the division by 500 by *multiplying* by 500:

$\tfrac{2}{5} \times 500 = \tfrac{x}{500} \times 500$, which yields just x on the right
$\tfrac{2 \times 500}{5} = x$
$200 = x$

As previously mentioned, a procedural technique called *cross-multiplying* can also be used, but is not recommended. As this procedure does not support understanding at the elementary level, we have opted not to provide the details of it here.

Alternately, a *ratio* or equivalent fractions-type technique can be used to solve the equation (see section 6.5).

$$\tfrac{2}{5} \dashrightarrow = \dashrightarrow \tfrac{x}{500}$$

Since the fractions are equal and 500 is 100 times as much as 5, then x has to be 100 times as much as 2. Therefore, x must be 2×100, or 200.

If the example gave us different initial information, such as how much Ben paid in the bicycle example, then we might have a different equation:

$$\tfrac{2}{5} = \tfrac{200}{x}$$

At first glance, this equation seems more difficult because the unknown is in the denominator position; however, by following the rules of solving equations, we can still perform the same operation on both sides. Multiplying both sides by 5 simplifies the left side and removes the fraction from the equation:

$$5 \times \frac{2}{5} = \frac{200}{x} \times 5$$
$$2 = \frac{1000}{x}$$

We can then multiply both sides by x, which is allowable even though we don't know its value (although we must remember that x cannot be zero). This gives us

$$2 \times x = \frac{1000}{x} \times x$$

Since dividing by x and then multiplying by the same x is 1 (or, informally, the "x's on the right side divide out") then

$2 \times x = 1000$, so x is 500.

A third method that can be used to solve the problem involves the same type of ratio reasoning as in the previous method for solving $\frac{2}{5} = \frac{200}{x}$. Since 200 is 100 times as much as 2 (comparing numerators), x must be 100 times as much as 5, or 500. A model that may help students see the relationship between these values is the ratio box. The relationship of the values can be seen as follows:

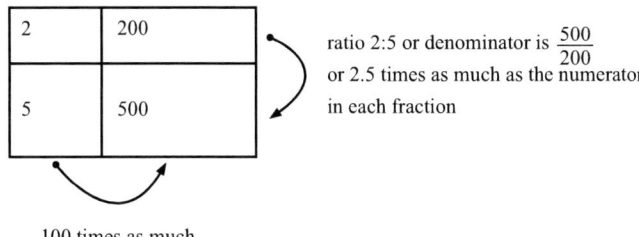

The ratio box connects ratios, fractions, and algebraic concepts, and is handy as both a solution tool and a way to make sense of both a problem and its answer.

Practice and Further Exploration

1. Use a ratio box to determine the relationship between coins and people if 12 coins are shared equally among 3 people. Label the rows and columns as to people and coins.
2. At the grocery store, you notice that one type of dog snack comes in 454-gram bags that cost $9.95 and 1.2-kilogram bags that cost $14.98.
    a) By estimating and using mental math, determine which is the better deal.
    b) What is the exact price per 100 grams in each case? (Feel free to use a calculator.)
    c) Referring to the methods illustrated in this section as a guide, use another method to determine the exact price per 100 grams in each case.

12.3 Building Patterns That Change in Two Dimensions

In the first two sections of this chapter, we have developed numeric methods for solving relationships or equations in which there is only one unknown value. We now turn to relationships that represent patterns, in which multiple input and output value combinations are possible.

The first step in exploring patterns of any type is to recognize how they are changing; this allows us to predict the next element based on the previous elements. Patterns can develop in many ways—for example, they can simply repeat themselves, as they do in a wallpaper pattern or on a tiled floor, or they can grow or shrink. The patterns described previously were those with a constant amount of change or a constant recursive rule or solution—for example, patterns in which a set number of lines or tiles were added to each new picture. But other patterns change in more complicated ways.

Patterns in the real world are often much more complicated than the linear patterns we have seen thus far. While they may still change predictably, the change is more complicated than simply adding a certain number each time. Consider the following pattern example, made of small squares, referred to as an *alge-animal*. The alge-animal is the main example that we will use throughout the rest of chapter.

> **MATHEMATICAL TERM**
>
> The *recursive rule* or solution for a relationship tells how much the output changes when the input increases by 1, from a given starting point; for example, *start with 2 and add 3 each time*.

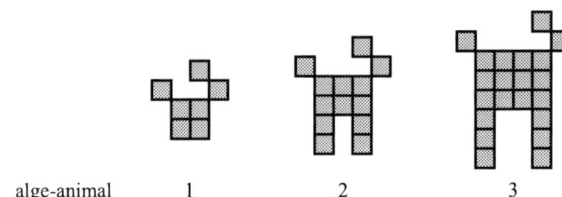

alge-animal   1   2   3

Exploration/Task

1. If the alge-animal continues to grow according to the pattern shown above, sketch the fourth alge-animal. As you draw, think about how knowing that it is the fourth picture helps you predict the number of squares in each part. Repeat for the fifth alge-animal.
2. The following sketch represents the tenth alge-animal. Indicate the dimensions of the body and legs on the diagram. Use this information to calculate the total number of tiles needed to build it. If you find this to be too difficult, return to your sketch of the fifth alge-animal. Think about how the number 5 helps determine the size of the parts. If necessary, draw the sixth picture.

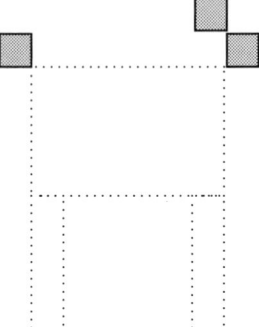

3. Use the reasoning you applied in the previous questions to determine the total number of tiles in the fiftieth alge-animal, assuming that the pattern keeps growing in the same way. Show your calculations.

Follow-Up and Discussion

The key to working towards seeing, and eventually predicting, resultant patterns is to explore the link between the picture number and whatever we are counting. In the following section, we will continue to explore this idea using the same example. Solutions to the above problems are described in detail in the following section.

12.4 Describing Patterns That Change in Two Dimensions

We see that the legs of the alge-animal grow in accordance with the picture number. For example, in the third picture each leg is made using three tiles, while in the tenth picture each leg is 10 tiles long; thus, if the pattern continues in the same way, in the fiftieth picture each leg would be 50 tiles long. Each time we build the next alge-animal in the pattern, we need 2 additional tiles to build the legs. Since the legs keep growing at the same rate, the growth is called *linear*.

On the other hand, with each new picture the body grows in *two* dimensions. (Conceptually, this is linked to the idea of measuring area, which will be discussed in chapter 14.) It is important to notice how each dimension changes, particularly in relation to the picture number. Focusing for the moment only on the body of the alge-animal, the following table provides the dimensions of the body for each picture:

Alge-animal	1	2	3	4	...	10	...	50
Height	1	2	3	4	...	10	...	50
Length	2	3	4	5	...	11	...	51
Body dimensions	1 × 2	2 × 3	3 × 4	4 × 5	...	10 × 11	...	50 × 51
Body area	= 2	= 6	= 12	= 20	...	= 110	...	=2,550
Tiles added		+ 4	+ 6	+ 8				

We can use the dimensions of each rectangular body part to calculate the number of tiles in it. When we look at the total tiles in each body part, we see that, unlike the legs, the number of tiles added to the body to create each new picture is not a constant. We say the growth is *non-linear*. This phrase comes from the appearance of the graph that can be produced to show this growth; as we will see in section 12.5, the graph showing the body size is not a straight line.

Lastly, looking back at the remaining parts of the alge-animal, we see that exactly 3 tiles are used for the tail, head, and ear in every picture. This part is *constant*—it does not change with each picture.

> **MATHEMATICAL TERM**
>
> A *non-linear* relationship is represented by a graph that is not a straight line.

> **MATHEMATICAL TERM**
>
> A *constant* is a name given to an unchanging amount; for example, the number 5 is a constant.

Using the above information, we can calculate that the total tiles needed to create the fiftieth alge-animal picture is: body + both legs + head, tail, ear = 50 × 51 + 2 × 50 + 3 = 2,550 + 100 + 3 = 2,653.

Practice and Further Exploration

1. Use the alge-animal example illustrated in this and the previous section to predict the number of tiles in the twentieth picture.
2. Using square tiles, create a different alge-animal that contains both parts that grow linearly and parts that grow non-linearly, as well as parts that are constant. Draw the first three pictures, then sketch the outline of the tenth picture and use the outline to calculate the number of tiles in the tenth picture.

12.5 Graphing Patterns That Change in Two Dimensions

One way to get a visual idea of how patterns change is to create a graph. A convention in mathematics is to use the horizontal axis of the graph for the picture number or the input number. (In more advanced mathematics, this value is referred to as the *independent variable*.) The result or answer, which is sometimes also called the output (the quantity we are trying to predict), is recorded on the vertical axis. (Since this value depends on where in the pattern we are, it is sometimes called the *dependent variable*.)

In higher level mathematics, a great deal of time is devoted to learning to create graphs based on knowing the pattern rule, relationship, or equation for the pattern. At the intermediate level, graphing relationships for patterns that change in two dimensions (such as the alge-animal) are often created using a table of values.

Exploration/Task

Complete the following table of values for the alge-animal example used in the previous sections. The first entry has been completed for you. See if you can find a pattern in the numbers in the second column to help you predict the last entry without using a picture to calculate the value. Hint: Continue to look for a pattern in how the number of tiles added is changing.

Picture number	Number of tiles
1	7
2	
3	
4	
5	
6	

Follow-Up and Discussion

You might also calculate the number of tiles for the sixth and seventh alge-animals by imagining the pattern in your mind. You might have reasoned something like this for the sixth one:

The body is 6 ×7, that's 42 tiles in the body. There are 2 legs of 6 tiles each, which is another 12. There are 3 more for the head, ear, and tail. In all, that's 42 + 12 + 3 is 57.

You may also have noticed that the number of tiles added to each picture to create the next picture changes at a constant rate; we keep adding 2 more than the last time. The first 6 output values are 7, 13, 21, 31, 43, and 57. Starting from 7, we added 6 to get 13. We then added 8, 10, 12, and finally 14 to get 57.

Exploration/Task

Using your table of values, create a suitable scale and graph the alge-animal pattern on grid paper.

Follow-Up and Discussion

Since the vertical axis must include values up to 57 (the number of tiles in the sixth picture), it makes sense to use a different scale for the vertical and horizontal axes. In the graph shown below, one vertical space represents 2 units, while one horizontal space represents 0.5 units (i.e., 2 spaces are used to represent each single unit).

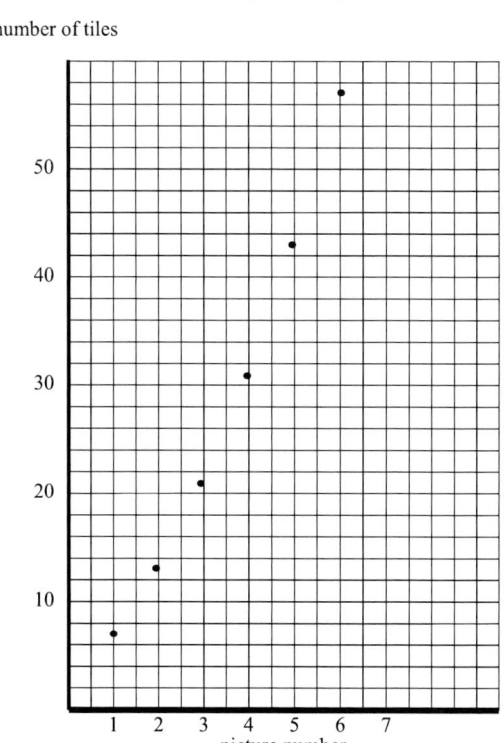

This type of graph is called a *scatter plot* as it contains points, rather than a line. Since the pattern was defined with whole number amounts of alge-animals, it doesn't make sense to include less-than-whole alge-animals on the graph. However, in a different context, if it made sense to have fractional or decimal input values, we could join the points with a smooth curve.

We can continue the graph for larger and larger picture numbers (sometimes called input numbers or term numbers). You might wonder if there is a point on the graph for a term number of zero. If we use the numeric calculation method for the number of tiles for a term number of zero, the result would be 3, giving the point (0,3). We can think of the 3 as the 3 constant tiles, or the alge-animal with zero growing parts. You can add the point (0,3) to your graph and observe how the addition of this point makes the curve of the graph more pronounced.

This relationship does not produce a straight-line graph, and this is where the name *non-linear* comes from. The graph is literally not linear. If we graphed the seventh alge-animal, we would see that the curve becomes even more pronounced, with 73 tiles needed. Imagine where the point (7,73) would be on your graph.

Continuing to look at the graph, we can see that as the picture number continues to increase, the output values of the points on the graph seem to increase at a greater and greater rate. This is different from the patterns explored in chapter 11, which changed by the same amount each time—and thus are called linear. We will explore this idea further in the next section.

Practice and Further Exploration

Explore the following pattern. Draw the next picture, assuming the pattern continues in the same manner.

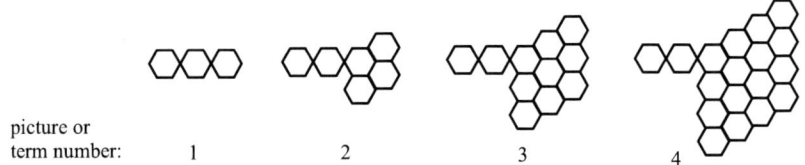

picture or term number: 1   2   3   4

1. Create a table of values, showing the picture or term number and the number of hexagons for at least the first five pictures. Can you find a way to determine the number of hexagons in the sixth and seventh pictures without drawing them? Find a rule that uses the term number to calculate the number of hexagons, and test it on a few terms.
2. Graph the term number and the number of hexagons as points on a graph. Think about what the zeroth term might look like. Calculate the number of hexagons for the zeroth term using the rule you developed. Add a point on the graph to represent this finding.

12.6 Rates of Change, Slope, and First Differences

In the middle elementary grades, patterning is generally addressed by drawing models. At the intermediate level, however, the ability to construct a general pattern rule becomes important. A pattern rule (often called a formal rule, or properly, a *function rule*) expresses the pattern outcome for *any* location or term in the pattern. For example, recall the growth of the legs of the alge-animal earlier in this chapter:

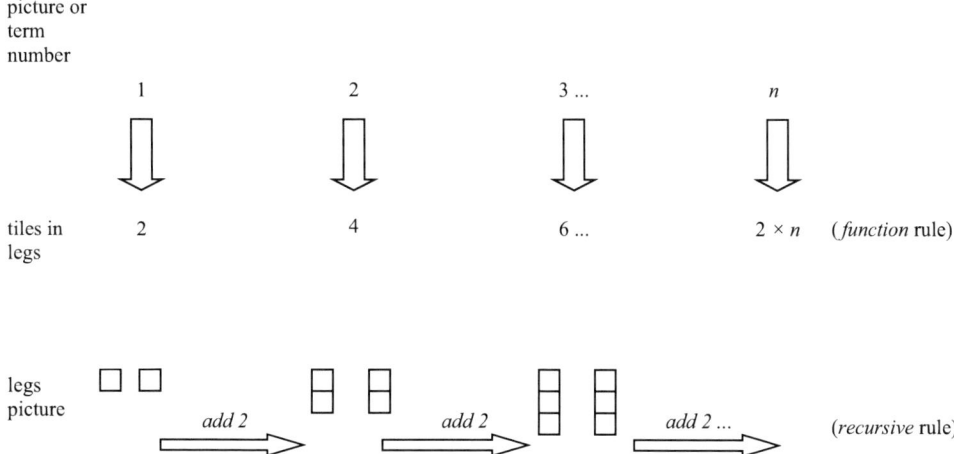

The function rule—the rule that uses the picture or term number to tell us the number of tiles in that picture in the pattern—for the legs is $2 \times n$ or $n \times 2$.

Note that to find the number of tiles in the legs in the fiftieth picture using a function rule, we do not need to calculate all the previous values. There is, however, a connection between the function rule and the recursive rule (recall this is the amount added to get the next picture or outcome)—the recursive rule is *start with 2 and add 2 tiles each time*. It makes sense that the number of times we add 2 is related to the picture number. You can also think of the function rule of $n \times 2$ as n is the number of times that 2 was added.

> **MATHEMATICAL TERM**
>
> The *function rule* in a relationship tells how to find the output or result for a particular input value or term number.

Exploration/Task

Examine the graph of the legs pattern in the alge-animal example, and look for connections between the graph, the function rule, and the recursive rule, as well as within the graph itself.

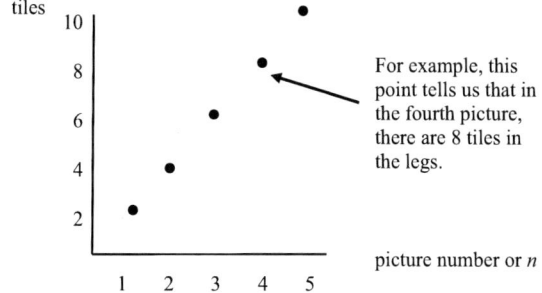

For example, this point tells us that in the fourth picture, there are 8 tiles in the legs.

Follow-Up and Discussion

You might have noticed the connections among the number of tiles added each time (the recursive rule), the 2 that multiplies the picture number (the formal rule), the *difference* between each output value, and even the way the line changes on the graph. All of these ideas are related—they are all connected to how the pattern is changing (in this case, growing). They are related to the *rate of change* of the graph. Examine the same data in a chart, which relates to the graph.

Picture number (n)	Number of tiles in legs (output or result)
1	2
2	4
3	6
4	8

+2
+2
+2

2 more tiles each time

The recursive rule tells us how the pattern is growing, as shown to the right of the chart. The recursive rule is also related to the *first difference*—which is particularly easy to identify when the data is shown in a chart form. Literally, the first difference is the difference between output values.

It makes sense that if the change implied by the recursive rule, or the first difference, is a constant value, the graph is a straight line (linear). Visually, the graph of a linear function rule changes by the same amount each time. The graph shows that the output—here the number of tiles in the legs of the alge-animal—increases by 2 with each new *n* value. The concept of rate of change is also related to the slope of the line. This same value, the 2 in our example, appears in the function rule of $2 \times n$.

> **MATHEMATICAL TERM**
>
> The *first difference* is the value added (or subtracted) each time according to the recursive pattern rule. It is the difference between output values when the input increases by 1.

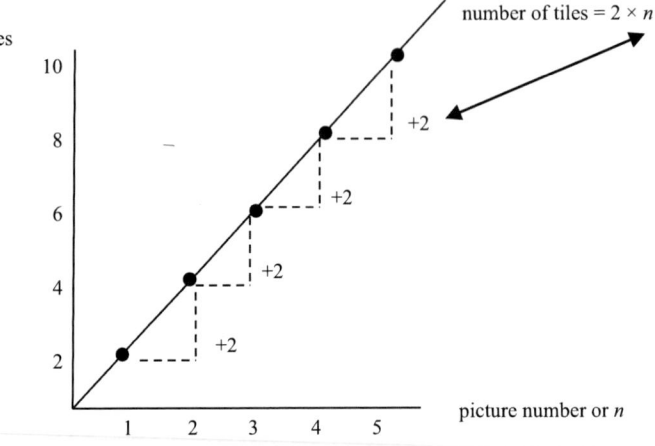

Exploration/Task

For the following pattern made of rectangles, predict the number of rectangles needed for each picture number, assuming the pattern continues to grow in the same way. Using the questions below as a guide, look for a connection among the ideas of the growth of the pattern, the recursive rule, the first difference, the function rule, the graph itself, and the slope of the graph.

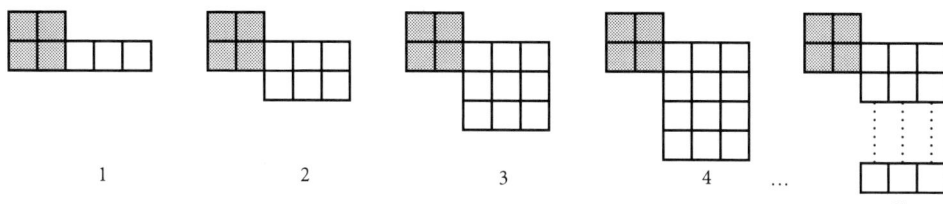

1. The recursive rule might be stated *start with 7 and add 3 each time*. Interpret this rule to determine how many rectangular tiles must be added to get from the fourth to the fifth picture.
2. What would term number zero look like?
3. Create a chart comparing the picture or term number to the number of rectangles for the first 5 pictures. What is the first difference?
4. Using your chart, graph the picture number (on the horizontal axis), and the number of rectangular tiles (on the vertical axis), up to the fifth picture. If the dots were joined, what would be the slope of the graph?
5. Compare the recursive solution, the first difference, and the slope of the graph. How do they relate?
6. In the pattern, there are 4 shaded tiles in each picture that don't change in number from picture to picture. How is this aspect shown in the graph?
7. State the function rule for the pattern. What does this rule tell us about the graph?

Follow-Up and Discussion

The recursive rule predicts that the pattern grows by 3 tiles each time; therefore, we need 3 more tiles for the next picture. In the zeroth picture, we would have only the initial 4 tiles. From there, each new output value would have 3 more tiles than the last. The first difference and the slope of the graph (if the points were joined) are 3, which agrees with the growth predicted by the recursive rule. All of these aspects tell us how the pattern is changing, so the phrase *rate of change* is sometimes used for these ideas. The 4 shaded tiles that appear in each picture relate to the zeroth term, and to the point (0,4) on the graph. In a linear function, this point, as well as the rate of change, are both found in the equation: number of tiles = 4 + 3 × n, where n is the picture number or term number. The larger the n value, the more times we have to add a set of 3 additional tiles. If we joined the points on the graph, the line would meet the vertical axis at (0,4). This point is sometimes called the *intercept of the line*; it is literally the place where the line touches or intersects the vertical axis (sometimes called the y-axis).

We return once again to the alge-animal example. We see that in addition to the leg parts, there are two other aspects to the alge-animal pattern: the head-ear-tail parts and the body. The head-ear-tail parts are the 3 tiles that appear once in each picture. If we consider this part of the pattern on its own, we see that the number of these tiles neither grows nor shrinks:

Picture number	Head-tail-ear
1	3
2	3
3	3

For this part of the pattern, all of the descriptions of the change according to the recursive rule, the first difference, and the slope are zero. In this case, we say that the rate of change is zero, and thus that this is a *constant* function. The number of these (head-tail-ear) tiles stays the same in each picture.

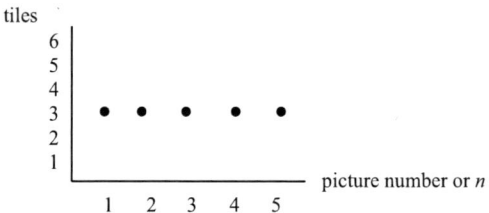

We see that a constant function, such as the head-tail-ear, has a horizontal graph. The number of tiles stays the same. The growth of the body of the alge-animal will be explored more formally in the next section.

12.7 Second-Degree Pattern Rules

The following pattern shows the increasing size of the body of the alge-animal:

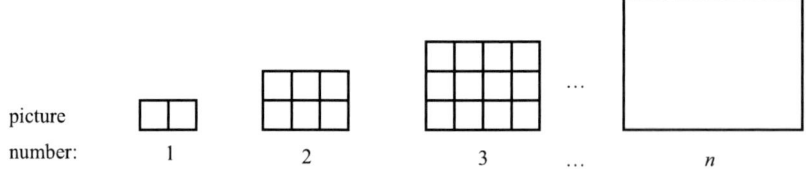

There are a number of ways of visualizing the picture in general, which we will call the nth picture, where n is the picture number. It is important to explore how the picture number determines the dimensions of the rectangle. Looking at the body, we might see a square and a rectangle, or simply a larger rectangle:

We might see the illustration on the left as describing an *n*-by-*n* square plus another length of *n*. On the right, we might see one rectangle with dimensions *n* by $n + 1$. Different ways of seeing the picture result in different-looking algebraic forms of the rule. The teacher must be on the lookout for these alternate ways of expressing the rule, both of which are equally correct.

Exploration/Task

Find two ways of algebraically expressing the function rule for the body of the alge-animal as suggested by the different pattern pictures of the body as shown using *n* in the dimensions.

Follow-Up and Discussion

You might have found the algebraic rules of $n \times n + n$ (which could have been written $n^2 + n$; we'll consider that the same rule as it relates to the same visual), for the "square plus a bit more" visualization, and also $n \times (n + 1)$ for the rectangle. Both are correct, and it is interesting for students to compare them and make sense of the different forms of the rule and how they apply to the picture.

Both versions of the rule involve the product of a length of *n* multiplied by *n*. Because each involves *two* factors of the same variable multiplied together (the $n \times n$ or n^2), we say that these are *second-degree* rules. Very roughly, the highest implied exponent determines the degree.

Exploration/Task

Complete the following chart for the alge-animal body. Determine the first difference between each number of tiles. Is it constant? Then construct a graph of the number of tiles in the body for the first five pictures.

Picture number	Number of tiles	First difference
1		
2		
3		
4		
5		

Follow-Up and Discussion

You should have found that not only was the first difference not a constant, the graph was not a straight line. The first differences (4, 6, 8...) do have a predictable pattern, however—they are a constant amount *apart*. If the first differences change by the same amount each time (in this case, a growth of 2 each time), then we say that the second difference is constant (in our case, 2). Interestingly, if the second difference is constant, the function will be second degree, which means that there will be a term that is second degree, or that has an exponent of 2 (for example, the $n \times n$ or n^2 term). The presence of an $n \times n$ term might be less apparent if we are working with the $n \times (n + 1)$ form of the rule, and students will need to understand how to expand brackets to simplify this version of the rule (this will be discussed in section 12.8). The key idea is that a second-degree function rule does not have a constant slope or rate of change—it either keeps on growing or shrinking faster and faster, or slower and slower. This is evidenced in the alge-animal body graph, which becomes increasingly steep as the graph is extended. Given that the entire alge-animal graph contains the (second-degree) body, it too, is non-linear as we have seen.

Picture number	Number of tiles	First difference	Second difference
1	2		
		}4	
2	6		}2
		}6	
3	12		}2
		}8	
4	20		}2
		}10	
5	30		

Second-degree (and any other non-linear) patterns are more difficult to work with, given that they don't have a constant recursive relationship or rate of change. It may be too difficult for students at the elementary level to find the pattern rules for questions that are not presented as a series of pictures. Visual exploration is important, and lays the groundwork for more formal study. Since most things in the world behave in a non-linear, rather than linear, manner, it is important for students to at least get a glimpse of these relationships.

The following illustration shows a well-known example of a combined linear and non-linear pattern that is accessible via visual representation. This problem is often called *the border problem*, because of the shape it creates. This problem is sometimes contextualized in situations using tiles around a square or other rectangular area, such as square patio stones surrounding a garden. Here is the pattern of square tiles:

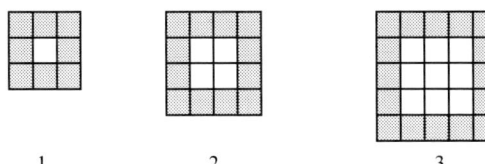

Picture   1   2   3

Exploration/Task

Familiarize yourself with the border problem (as shown above) by exploring it using the methods of this chapter. You may choose to explore the number of tiles in the complete picture for each term number, or the pattern of the white tiles and the grey tiles separately. If possible, find more than one way to count the tiles each time. You might explore and compare the recursive rule, the graph, the first and second differences, and the function rule for each.

Follow-Up and Discussion

In your exploration, if you considered the grey tiles or border separately, you might have found that the number increases by 4 tiles each time. This constant number that is added each time means the graph is linear. If the points were joined, the graph would have a slope of 4—the 4 represents the number of tiles added each time or the rate of change. Depending on how you grouped and counted the tiles in the border, in particular how you counted the corners, the pattern rule could be written as any of the following:

    a) $4 \times n + 4$
    b) $2 \times (n + 2) + 2 \times n$
    c) $4 \times (n + 1)$
    d) $4 \times (n + 2) - 4$
    e) $(n + 2)^2 - n^2$

In our teaching, we often find that students will construct a variety of rules. Again, we emphasize how important, but challenging, it is for the teacher to get inside their students' minds to understand their reasoning. For an interesting video of students working with a similar problem, and describing their thinking, see Boaler and Humphreys (2005). It would be useful to explore each of the pattern rules above (especially those you did not initially think of yourself), and the thinking they demonstrate, to see how the rule counts the tiles. The last rule looks like it might be a non-linear rule—yet the graph is still a straight line. Can you see why?

In the next section, the rules of algebra will be developed using a new set of manipulatives, which will allow students to determine that the rule $(n + 2)^2 - n^2$ contains an n^2 term but also a "$-n^2$" term. In other words, the n-squared terms subtract away from each other in the equation, so it really is a linear rule. If you were not able to determine the idea behind this version of the rule, it might help to think of *the entire square with the white inside removed.*

The white tiles inside can be counted by seeing an n-by-n square, or n^2. Therefore, a rule for each entire shape (grey and white tiles combined) could be obtained by adding n^2 to any of the previous rules for finding the number of grey tiles only. Other rules for the entire shape can be determined by seeing the shape as a square that is 2 units larger than n in each dimension; so there are $(n + 2) \times (n + 2)$ tiles in all; or $(n + 2)^2$. There are other ways to construct and write the rule. In section 12.8, algebra tiles will be explored as a way to resolve that these rules are algebraically the same as our other versions of the rules.

Practice and Further Exploration

For the following pattern, the first three pictures are shown.

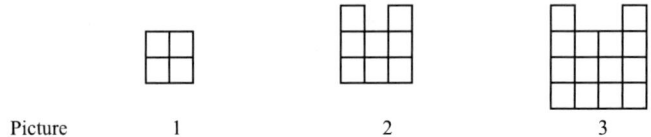

Picture   1   2   3

Assuming the pattern continues as illustrated:

1. Draw the next picture.
2. State the function rule, then work to find at least one more possible version of the rule using a different way of grouping and counting the regions. By using circles, arrows, and so on, show how each algebraic part of each rule relates to a picture of a term in the pattern.
3. Create an input-output chart for the first five terms and explore the first and second differences.
4. Graph the pattern for at least the first five terms, comparing the picture number to the number of tiles.
5. What do the first and second differences tell us about the graph?

12.8 From Algebra Tiles to Algebraic Methods

Algebra tiles are a manipulative that can be used to represent numbers, as well as variable amounts. They have a resemblance to base ten blocks, but include representations of lengths and areas that *vary*—the key additional feature of algebra tiles is that they contain variable pieces. While the number 1 is still represented by a unit cube, some additional pieces are included in algebra tiles. The best way to visualize algebra tiles is to see them dynamically at work, such as on the National Library of Virtual Manipulatives site (http://nlvm.usu.edu/en/nav/category_g_3_t_2.html). Choosing a variable piece (which we can call n, x, y, or another term) and dragging it onto the working area, then using the mouse to change its size, reinforces the important idea that the variable pieces are indeed designed to represent amounts that vary. The variable pieces represent a general value—any value—rather than a specific number.

Using the examples from the alge-animal body explored earlier, students might wonder how to find out if $n^2 + n$ and $n \times (n + 1)$ really do represent the same amount. To explore this, we will designate a variable piece as n. We note that it is a plastic rod with a width of 1, but it must be understood as representing a variable length, denoted by n. It can be thought of as the sum of n unit pieces, lined up to form a rod. Its width is 1, and its length is n.

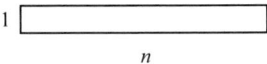

Similarly, we can use a *square* of dimensions n to represent n squared, or n^2; n^2 is literally an n-by-n square. A key point is that as n changes, so does n^2. Since both dimensions of the square are n, the square grows in both directions, and stays square. Once again, this is easier to see with a virtual, rather than plastic or drawn, representation.

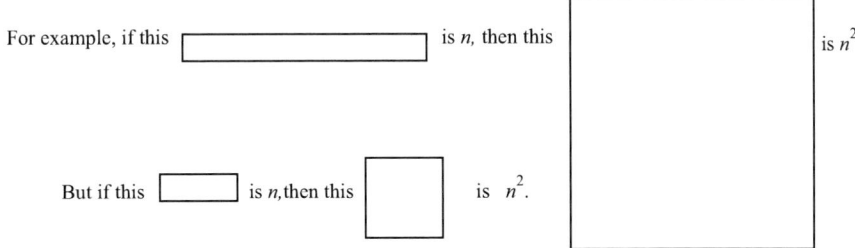

Exploration/Task

Use algebra tiles to explore $n^2 + n$ and $n \times (n + 1)$. Model both expressions using algebra tiles and compare them. Recall the idea of using an area model to represent an expression such as $n \times (n + 1)$: the two parts or factors form the dimensions of the rectangle, and the answer or product is the area enclosed.

> **CONNECTION**
>
> To review multiplication models such as the area model, see section 5.2.

Follow-Up and Discussion

Using the idea that n^2 is a square, we can model $n^2 + n$ by combining an n piece with an n^2 piece.

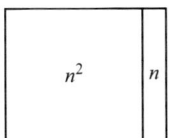

We recall that the n piece is really a 1-by-n rectangle. We can mark the dimensions on our model as follows:

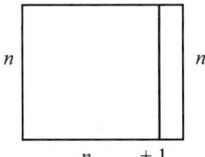

We started with a model of the expression $n^2 + n$. Labelling the dimensions of the two algebra tile pieces makes it clear that the same model also shows $n \times (n + 1)$

—the two expressions are the same! Algebra tiles have the neat capacity to help students literally construct algebraic rules. In the following task, you will construct yet another rule that you may recognize. After completing the task, you won't need to use the rule, because you will simply understand what is happening.

Exploration/Task

Using the idea that a product can be expressed as an area, construct a rectangular area to represent $(x + 2) \times (x + 3)$. Remember that when using an area model, the two factors each represent side length, and the product is the area of the rectangle (see section 5.2). Examine the resulting area to find another way to write the product. The previous discussion may help.

Follow-Up and Discussion

A slightly easier version of the previous task would be to ask students to determine if the following algebra tiles pieces can be arranged to form a rectangle and, if so, what the dimensions of the rectangle are. This latter task is in fact a *factoring* task, while the originally proposed task is a *simplifying* task. The two tasks are really the same task with different starting points; we will discuss the factoring task first, and then return to the original problem. Here are the pieces that need to be arranged:

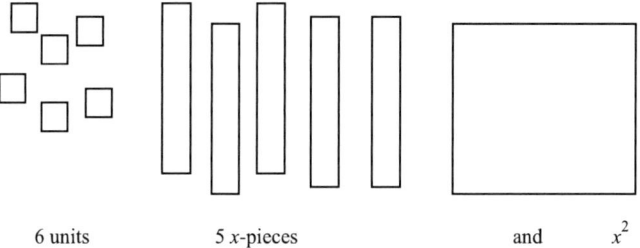

6 units   5 *x*-pieces   and x^2

It is important to remember is that the *x* pieces are variable in length. Consequently, we can't line up units along an *x* piece, because we don't know how long it is.

The *x*-piece might be this long

or this long

After some trial and error, we might arrange the pieces as follows to construct a rectangle:

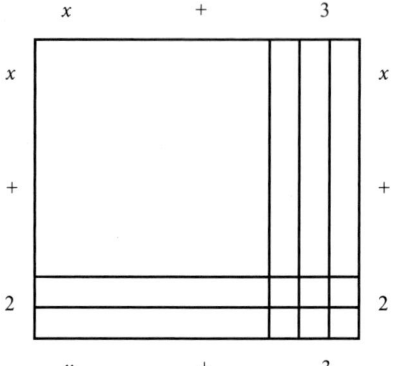

The ones, shown as unit tiles, can only be lined up against the ends (units) of the *x*-tiles, as these are the only parts of the *x*-tiles that do not change. The *x* dimensions cannot be predicted, so we do not know how many units to use along the *x* lengths.

We see that the rectangle made with the algebra tile pieces representing $x^2 + 5x + 6$ has dimensions $(x + 2)$ and $(x + 3)$. These dimensions are the factors. Similarly, the rectangle with dimensions of $(x + 2) \times (x + 3)$ can be seen to simplify to $x^2 + 3x + 2x + 6$, or $x^2 + 5x + 6$. Once this relationship becomes visually apparent, rules for performing such steps become things of the past. (You might recall being taught rules such as FOIL—"first, outside, inside, last"—which we do not recommend. And if you didn't remember what that stands for, it doesn't matter—it just illustrates how easily these rules are forgotten).

Students should use algebra tiles for as long as is needed to develop mental images of the operations. The procedures will gradually become apparent and internalized, after which students will no longer need the manipulatives. There are also integer versions of these algebra tiles, but we recommend using the positive versions as starting points for helping students initially develop early algebra concepts. Nearly of all the basic rules of factoring and simplifying expressions become apparent with algebra tiles or their virtual counterparts. Use either as you work through the following practice questions.

Practice and Further Exploration

1. Draw the following algebra tile pieces: y^2, 6 of the *y* pieces, and 9 units. Use trial and error to arrange them into a rectangular shape. Use the dimensions of the rectangle to determine the two factors that multiply together to give $y^2 + 6y + 9$. (Hint: Remember that squares are rectangles too!)
2. The expression Problem 1 is called a *perfect square*. In more advanced algebra, students learn rules for determining if an expression is a perfect square. What is the relationship of the values in this expression that makes the model a square shape?
3. Use algebra tiles to model the expression $(x + 5) \times (x + 2)$. Use your model to find the simplified expression.

12.9 Exponents as Multi-Dimensional Numbers

We have already explored the idea that if a variable such as *y* represents a length, then *y* multiplied by *y*, or y^2 can be represented geometrically as a square. What if we multiply by *y* again? Three of the same factor multiplied together, such as *y* times *y* times *y*, can be written y^3—literally, it is a cube. We recall that the thousands cube or block in base ten blocks is a 10-by-10-by-10 cube, and the number of unit cubes it contains can also be written as 10^3. Similarly, y^3 is a cube of dimensions *y* by *y* by *y*.

What about y^4? Since we live in a three-dimensional world, it is difficult to model four dimensions. At best, we can try to model a four-dimensional object with

> **CONNECTION**
>
> To review the use of base ten block manipulatives to model 10, 100, and 1,000, see section 3.3.

a three-dimensional representation. This is a bit like drawing a two-dimensional paper sketch of a three-dimensional cube. When we begin to deal with expressions such as y^4 in mathematics, our concrete models are less helpful. Knowing what the exponent is really doing is important, particularly when designing interventions for students who are confused about the difference between $2 \times x$ and x^2.

Practice and Further Exploration

1. Sketch a paper representation of $3 \times n$ and n^3. Use the idea of algebra tiles, combined with your knowledge of base ten blocks, to come up with a meaningful representation of each.
2. Evaluate $3 \times n$ and n^3 for the value $n = 100$. You might want to think about how the n^3 could help us convert a cubic metre to cubic centimetres.

12.10 Exponentials

In the previous section we looked at multiplying by a variable, such as y, a certain number of times. In this section, we will look at a context in which we multiply by a constant value—but a *variable* number of times.

Exploration/Task

Explore the following scenario. As a birthday gift, you are presented with two gift options: you can receive $2 today, $4 tomorrow, $8 the next day, and so on, or you can choose to receive $100 today. How many days must pass for the first option to be better than the second?

Follow-Up and Discussion

In the first scenario, the amount received each day doubles, or is being multiplied by 2. On the third day, the amount is $2 \times 2 \times 2$ or 2^3. The amount on any given day d is 2^d. This is called an *exponential function*. The variable is the exponent, and it tells us how many times we are multiplying by the value. If the value being multiplied—the *base*—is greater than 1, the function keeps getting bigger. On day six, for example, the amount received according to the first option is 2^6 or $64. It doesn't take very long for the first option to be a better deal than the second!

Exponentials have many physical applications, such as bacterial growth or radioactive decay (in which case the base is less than 1). Since population growth is exponential, these functions are important to understanding the future of our planet.

Practice and Further Exploration

1. The amount of smoke remaining in a forest after a fire is 50% or half of what it was the previous day. (You can think about this as multiplying by

0.5 for each day). How many days must pass before the amount is less than 10% or 0.1 of the original amount?

2. The population in a fast-growing city is increasing by 20% each year, so next year the population will be 1.20 times what it is today. If the population growth continues at this rate, how long will it take the population of 100,000 to grow to 300,000?

Chapter Problems

1. Imagine a pattern made by arranging round counters in rectangular arrays. The first shape is a rectangular shape of 2 counters by 4 counters (8 counters are used in all), the second is 3 by 5, and the third is 4 by 6.
   a) Sketch these first three pictures in the pattern.
   b) One student told you that the rule was $(n + 1) \times (n + 3)$. Test this rule by sketching the fourth picture and determining how the rule connects to the sketch. Use the third picture to double-check the rule.
   c) Another student told you that the rule was $n^2 + 4 \times n + 3$. Test this second rule by sketching the fourth picture and determining how the rule connects to this sketch. Use the third picture to double-check the rule.
   d) Use algebra tiles to model each student's rule. How do the two rules compare?
2. A pattern is constructed of circular discs arranged in square arrays. The first picture is a 3-by-3 square, the second is a 4-by-4 square, and the third is a 5-by-5 square. If the pattern continues to grow in this way, find an algebraic rule using a picture number of n to calculate the number of discs needed to build the nth picture. Illustrate how your rule connects to the fourth picture.
3. Find a different way of grouping and counting the discs described in Problem 2 that yields a different-looking algebraic rule. Again, illustrate how the new rule connects to the fourth picture.

Further Reading

Anderson, K. L. (2012). Pattern-block frenzy. *Teaching Children Mathematics,19*(2), 116–121.

Anderson details lessons about ratios using online pattern blocks for Grade 6 students.

Collins, A., and Dacey, L. (2011). *The Xs and whys of algebra: Key ideas and common misconceptions*. Portland, ME: Stenhouse Publishers.

Collins and Dacey use modules in a flipchart to explore foundational concepts in algebra. This publication contains the mathematics knowledge and instructional resources necessary to teach the concepts students need to understand.

Kajander, A., Fredrickson, E., Casasola, M., and Boland, T. (2013). "Does anyone have another way?": Patterning, algebra, and inquiry in the elementary classroom. *OAME/AOEM Gazette, 51*(3), 28–34.

> This article illustrates the value of multiple versions of pattern rules in the development of understanding.

Smith, T. M., Seshaiyer, P., Peixoto, N., Suh, J. M., Bagshaw, G., and Collins, L. K. (2013). Exploring slope with stairs and steps. *Mathematics Teaching in the Middle School, 18*(6), 370–377.

> This article presents real-life context activities to help students develop a concept of slope and rate of change. The activities allow students to explore the concepts through a science and engineering context.

Stacey, K., and MacGregor, M. (2000). Learning the algebraic method of solving problems. *Journal of Mathematical Behavior, 18*(2), 149–167.

> Stacey and MacGregor detail a study conducted to examine student thinking and response to algebra.

Chapter 13
Geometry

13.1 Terminology

Geometry is one of the oldest branches of mathematics. The word *geometry* is connected to earlier forms of the words *geography* and *measurement*; historically, geometry provided a way to explore our physical surroundings using mathematical methods, but it has evolved to very abstract forms. There is new terminology introduced in geometry, and while some terms can be connected to everyday language, others cannot. Confusion can also arise with geometric terms that are used slightly differently in everyday language than they are in mathematics; for example, in everyday use the term *rectangular* usually refers to a shape that is *longer* or wider in one dimension than the other, which is certainly not a mathematical requirement of rectangles. In teaching, we suggest using concrete manipulatives and vocabulary walls with pictures that students can refer to, rather than memorizing terms. With modelling and enough practice, students will internalize the proper vocabulary.

13.2 Straight Lines and Angles

Straight lines are one dimensional, which means that they have only one measurement, namely, a length. We can't really draw a straight line, since it technically has no thickness. If we try, we end up with a representation that has at least a bit of width, such as a width of a pencil line; hence, we can only create the *illusion* of a straight line on paper. This idea makes for an interesting classroom discussion in which students are encouraged to come to terms with mathematical inaccuracies in models and representations. In other words, we might draw a straight line that has a tiny width using pencil or chalk, but we agree that our drawing *represents* an object with *no* measurable width.

If we have more than one line on a flat surface, the lines will either be parallel or they will eventually intersect, but we would have to extend them far enough to see where they meet. Geometric interpretations based on the idea that parallel lines do not meet are called Euclidean. There are other geometries (for example, what if space was curved instead of flat?), but they are beyond the scope of this book.

Line segments are lines with definite end points; in practice, we have to represent all lines as line segments, but mathematically, the term *line* refers to a straight line that extends infinitely in both directions.

A number of relationships exist with respect to lines and angles on a *plane*.

> **MATHEMATICAL TERM**
>
> A *line* is a one-dimensional object that extends infinitely in both directions. A *line segment* has a measurable length; in other words, it is like a line, but has a definite beginning and end point.

> **MATHEMATICAL TERM**
>
> A *plane* is a flat or two-dimensional surface.

196 Chapter 13

> **MATHEMATICAL TERM**
>
> *Parallel* refers to surfaces (or lines) that do not ever meet.

Parallel line segments, for example, will always be the same distance apart, as long as we are measuring the shortest distance between the two. This is the distance measured *perpendicular*, or at right angles, to both line segments.

Other properties of straight lines can and should be determined through exploration. Software and online environments offer the best place to do this. (This will be discussed in more detail later in the chapter.) When straight lines intersect, they meet at an angle. An angle is a measure of rotation; 360 degrees is a full rotation. The use of the number 360 as a full rotation may have derived historically from an estimate of the number of days it takes the Earth to orbit the sun.

> **MATHEMATICAL TERM**
>
> A *perpendicular line* or line segment is one that meets another line or segment at a 90-degree angle, sometimes called a *right angle*.

Exploration/Task

Explore the relationships of the labelled angles as shown. The arrow symbol indicates parallel lines.

1. Compare angle *a* to angle *b* in each diagram. These are called opposite angles.

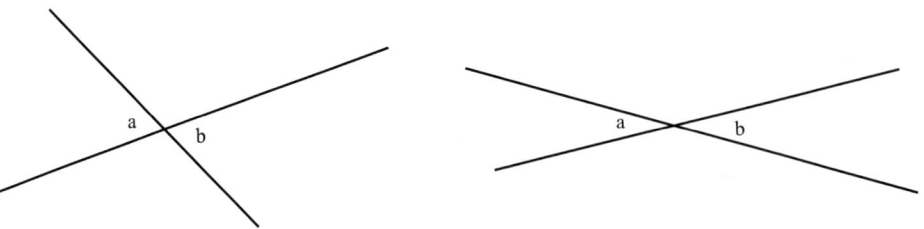

> **MATHEMATICAL TERM**
>
> A *degree* is a unit for measuring the angle of rotation. A complete rotation, such as a clock hand rotating around the face of the clock exactly once, is measured as 360 degrees. Half a rotation is 180 degrees, and so on.

What might you conjecture about the relationship all such opposite angles *a* and *b* have to each other?

2. Compare angles *a* and *b*, and angles *c* and *d* in the following diagram. Which pairs are equal? Construct another similar diagram (using two parallel lines and a third line that cuts them both, called a *transversal*), and explore whether the relationships you conjectured among the four angles will always hold.

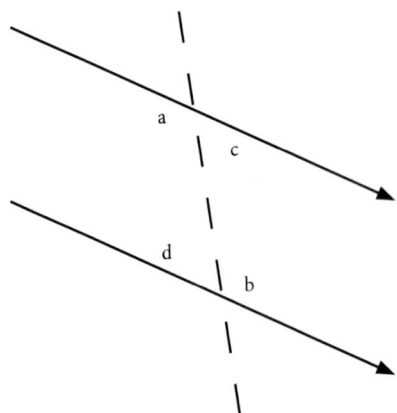

> **MATHEMATICAL TERM**
>
> A *transversal* is a straight line or segment that crosses or intersects two or more parallel lines.

> **MATHEMATICAL TERM**
>
> Two angles that add up to 180 degrees are called *supplementary angles*.

3. Using the above diagram, explore the sum of angles *a* + *c* and *d* + *b*. What does knowing these two sums tell us about *a* + *d* and *c* + *b*, given what you found in Problem 1?

Follow-Up and Discussion

The relationships explored above are called the *parallel line theorems*. Opposite angles formed by two intersecting lines are equal, and angles that add up to a straight line must total 180 degrees (half of 360 degrees). These relationships can be helpful if some—but not all—of the required angle measures are available to us, and we want to determine the missing angle measures. In Problem 1, a and b are equal. In Problem 2, you might have conjectured, by visual inspection, that $a = b$ and $c = d$. In Problem 3, both $a + c$ and $d + b$ total 180 degrees. Since the lines are parallel, it is also true that $a + d$ and $c + b$ total 180 degrees; we can deduce more formally that $a = b$ and $c = d$, as conjectured in Problem 2.

Practice and Further Exploration

1. Draw any large two-dimensional shape on a large sheet of paper, or trace one on the ground in the sand or snow. Stand on a marked spot on the shape, and notice which way you are facing. Now walk around the shape. What angle have you turned in total as you walk around the shape? If you are not sure, try it again, paying particular attention each time you turn. (Hint: If you end exactly where you started, and at some point in your walk you faced in each direction of a full rotation, how much have you turned in all?)
2. Imagine that you are placing a border around a shape, but you cannot see the fourth corner, as shown below:

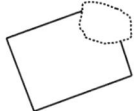

   How could you determine if the hidden angle was a 90-degree angle? If you need more information, what do you need to know?

13.3 Two-Dimensional Shapes

A two-dimensional shape in mathematics has exactly two dimensions—but not a third. Theoretically, a two-dimensional shape has no thickness. While we might cut out representations of two-dimensional shapes from paper, these are in fact three-dimensional shapes in that the paper has some thickness. (If it didn't, a stack of 100 sheets of paper would not have any height.) A *closed shape* has a definite inside and outside.

> **MATHEMATICAL TERM**
>
> A *closed shape* has edges that meet or a boundary with no gaps, such as a circle.

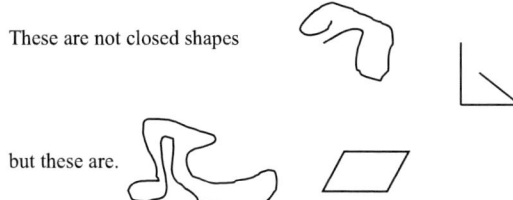

If we walk around the edge or outer boundary of any closed shape, we will have turned a total of 360 degrees. (If you didn't complete Problem 1 in the previous section, try it now! This is a great activity for children to do outside in the schoolyard; they can use chalk to draw shapes on the pavement. An observer can watch to see if the person walking around the shape has rotated exactly once in total.)

Shapes can have curved boundaries, such as circles, or straight sides (edges). We call shapes with straight sides or edges *polygons*, meaning *many sides*. *Regular polygons* have sides that are all equal in length. We need to be careful not to use the word *polygon* alone to imply *regular polygon*. For example, rectangles and parallelograms are polygons, but are not necessarily regular polygons. Squares (which are very special rectangles) *are* regular polygons. In fact, a square is a *regular* rectangle—it is a rectangle with all sides equal.

The following chart can be used to help construct a word wall or a vocabulary list for students to keep handy. Again, we suggest providing this list as a resource rather than asking students to memorize these terms. While all of the polygons with between 3 and 12 sides have specific names, we have included only the most common ones here. It should also be noted that it is not only important to include pictures, but that it is also important that the pictures represent several variations of each example. For example, students who have only seen models of triangles with a flat base and two other sides forming an upward point often mistakenly refer to triangles with a flat top and a two other sides forming a downward point as an upside-down triangle. As teachers, we need to ensure that we model shapes in a variety of positions. Similarly, examples of rectangles should include a square rectangle, since a square is a special case of a rectangle, namely the one with four equal sides.

> **MATHEMATICAL TERM**
>
> A *polygon* is a two-dimensional shape with straight lines as sides or edges. A *regular polygon* is a polygon with edges that are all equal in length. A square is a regular polygon, and a (non-square) rectangle is a polygon.

> **MATHEMATICAL TERM**
>
> The *perimeter* of a shape is its boundary or the measure of its boundary.

> **MATHEMATICAL TERM**
>
> Straight lines defining the outside of a closed two-dimensional shape are called *edges*. They are sometimes informally referred to as *sides*.

> **MATHEMATICAL TERM**
>
> *Vertices* are the points at which edges of a shape meet. For example, each adjacent pair of edges of a rectangle meet at a *vertex*, and a rectangle has four such vertices.

Two-Dimensional Shapes

triangle	3 sides	
quadrilateral (rectangles, parallelograms, trapezoids, rhombi, and squares are all quadrilaterals)	4 sides	
pentagon	5 sides	
hexagon	6 sides	
octagon	8 sides	
decagon	10 sides	

The *perimeter* of a shape is its boundary or the measure of its boundary. Some shapes have *edges* (straight lines) for the boundary, while others, such as circles, have curved perimeters. Edges meet at points called *vertices* (and each such point is a *vertex*).

Quadrilaterals, in particular, have many special names depending on their attributes. Venn diagrams are sometimes used to describe these characteristics, and to show how some shapes have more stringent classifications than others. For example, parallelograms have opposite sides that are parallel. That is also true of rectangles, but rectangles also have 90-degree angles, which may not be true of other parallelograms.

Exploration/Task

The following chart describes attributes of various classifications of four-sided shapes (quadrilaterals). For example, the trapezoid has two parallel sides, but the other attributes in the chart are not true of trapezoids. Think about these relationships as you complete the rest of the chart. Check a box only if it is true for all examples of that shape. The first line of the chart is completed for you.

Type of quadrilateral	One pair of parallel sides	A second pair of parallel sides	90-degree angles	Opposite sides equal	All sides equal
trapezoid	✓				
parallelogram					
rectangle					
rhombus (diamond)					
square					

Using the chart, we can see, for example, that all squares are rhombuses (or rhombi, the proper plural) because squares have all of the rhombus properties. However, not all rhombi are square ones.

Follow-Up and Discussion

Compare the completed shape chart to your own.

Type of quadrilateral	One pair of parallel sides	A second pair of parallel sides	90-degree angles	Opposite sides equal	All sides equal
trapezoid	✓				
parallelogram	✓	✓		✓	
rectangle	✓	✓	✓	✓	
rhombus (diamond)	✓	✓		✓	✓
square	✓	✓	✓	✓	✓

Practice and Further Exploration

1. Think through the relationships of four-sided shapes explored in this section and draw:
   a) a parallelogram that is not a rectangle, and is not a rhombus.
   b) a trapezoid that is not a square, and then one that is. Could either or both be a rectangle?
2. Explore the Illuminations activities on the National Council of Teachers of Mathematics website, in particular the Dynamic Paper activity (available at http://illuminations.nctm.org/ActivityDetail.aspx?ID=205). Use this tool to create some drawings of your choice.

13.4 Three-Dimensional Shapes

We live in a three-dimensional world. What would a two-dimensional being living on the plane of our floor see of us? What would it see when we stand up? What if we sat down and put our feet up on another chair? What would people living in a four-dimensional world see of us? Such questions and scenarios are interesting for children to consider, or even write a story about.

Almost all of the objects in our world are three-dimensional, and it is important to remember that children arrive in school naturally thinking in three dimensions. The ability to think in two dimensions develops as a result of doing a great deal of work with paper and pencil; it is not a natural development. As mentioned previously, even a piece of paper has a thickness—it is just very small. If we stack up a thousand sheets of paper, the stack would certainly be a number of centimetres high, which provides clear evidence of the thickness of each piece. Only in mathematics can we truly imagine a two-dimensional object such as a rectangle; in our world, the best we can do is *represent* the rectangle on a flat surface with a thin layer of ink.

Rather than working to reintroduce spatial reasoning to children after they lose their facility for it, which we have observed in some children by mid-elementary school, it is better to keep spatial reasoning as an intrinsic part of mathematics. Other subjects, such as art and science, can also provide opportunities to encourage spatial reasoning.

Just as two-dimensional shapes can be rearranged to form other shapes, we can also rearrange three-dimensional shapes. There are also different ways that we can view a three-dimensional shape; for example, a triangular prism (a *prism* is a shape that has two *faces* that are the same size and are parallel), can lie on a number of its faces:

> **MATHEMATICAL TERM**
>
> A *face* of a three-dimensional object is a flat surface on its outer surface. A cube, for example, has six faces.

> **MATHEMATICAL TERM**
>
> A *prism* is a three-dimensional shape that has at least two parallel faces that are the same size.

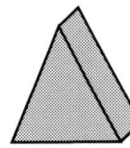 or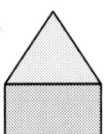

In the next chapter, we will see how some orientations are more helpful than others when measuring the volume of a shape.

As with two-dimensional shapes, there are a number of names for different three-dimensional shapes. Again, we suggest it is more important for students to focus on practising three-dimensional visualization than memorizing the vocabulary. A number of websites can help students learn about three-dimensional shapes; for example, the Illuminations: Resources for Teaching Math section of the National Council of Teachers of Mathematics site provides many useful activities. The Geometric Solids Activity for Grades 6 to 8 (http://illuminations.nctm.org/ActivityDetail.aspx?ID=70) allows students to change the sizes of shapes and rotate them in perspective. Students often experience more success with such activities if they have first had several opportunities to manipulate and explore concrete shapes—including both manipulatives and shapes that are found in the real world.

The *Platonic solids*, named after the mathematician Plato, are the three-dimensional objects that can be made using the same two-dimensional shape for its faces, placed at inside angles of less than 180 degrees (meaning that the shape is concave, like a soccer ball). Each of these shapes is also called a *polyhedron*. Examples include the cube (made with six squares), the tetrahedron (made with four triangles), and the dodecahedron (made with twelve pentagons).

The NCTM's Geometric Solids Activity cited above illustrates these Platonic solids and other shapes, and explores the relationship among the numbers of faces, edges, and vertices of these shapes.

Students can examine and learn about the relationship among faces, edges, and vertices fairly easily by looking at concrete samples of solids and making a chart. They will find that the number of faces and vertices added together is always 2 more than the number of edges, i.e., $f + v = e + 2$. Note that this activity also strongly connects to patterning, and, of course, requires reasoning. It is important to give students the opportunity to discover these relationships, rather than simply telling them about them. Of course, not all three-dimensional shapes have faces; a sphere, for example, has a curved surface. Encourage students to take on the challenge of looking for three-dimensional shapes in their daily lives.

> **MATHEMATICAL TERM**
>
> A three-dimensional object is *concave* if it has faces that meet with interior angles of less than 180 degrees; for example, a soccer ball is concave, while a three-dimensional star is not.

> **MATHEMATICAL TERM**
>
> The *edge* of a three-dimensional object is a line at which two faces meet.

Practice and Further Exploration

1. Determine the relationship among the number of faces, edges, and vertices of *polyhedra* (plural of *polyhedron*) by examining three-dimensional objects (concretely or on the Web) and listing the number of edges, faces, and vertices in a chart. Investigate at least three shapes.
2. Practice creating three-dimensional drawings on dot paper using the Isometric Drawing Tool on the National Council of Teachers of Mathematics website (http://illuminations.nctm.org/ActivityDetail.aspx?ID=125).

13.5 Similarity and Equivalence

Imagine each of the following scenarios, and compare the two related images in each case:

- You look at an image on your computer screen and then zoom in, or enlarge the image.
- You have two copies of the same photograph in front of you: one is 5 by 7, and the other is 10 by 14.

> **MATHEMATICAL TERM**
>
> *Similar* objects are different sizes, but all angle measures are preserved. We say that such shapes are *proportional*.

> **MATHEMATICAL TERM**
>
> If two shapes are *proportional*, then the side-lengths of each are in the same ratio.

> **CONNECTION**
>
> To review ratio and proportion, see section 12.2.

> **MATHEMATICAL TERM**
>
> *Equivalent shapes* are the same in every way except location.

- You place an object on an overhead projector or document camera, and then change the position of the lens to make the projected object larger or smaller.

In each case, the pairs of images are *similar*. This means, informally, that they look the same, although they are different sizes. All of the angles within the images remain the same regardless of the size.

If the side-lengths of similar polygons are proportional, we can use the dimensions of one to determine the dimensions of the other. For example, if we have two similar triangles, one with the dimensions of 3, 4, and 5 centimetres, and the other whose first side is 6 centimetres, then we can determine that the other two sides will be 8 and 10 centimetres in length.

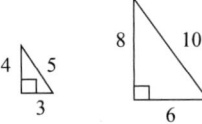

Shapes that are *equivalent* are the same, meaning that they are identical in every way, with the exception of location. Here are two pairs of equivalent shapes:

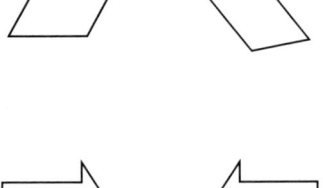

Knowing that a given shape is equivalent to another shape allows us to determine aspects and measurements of that shape by using its equivalent shape.

Practice and Further Exploration

1. Given that these two shapes are similar, determine length *a*.

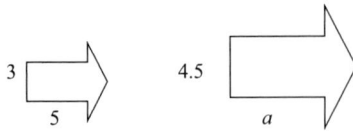

2. Given that these two shapes are equivalent rhombi, find length *b*. Explain how you know that this is the length.

13.6 Transformations and Symmetry

A transformation is the result of the movement of a shape to a new position or location. These movements are initially called *slides*, *turns*, and *flips*. At the elementary-school level, transformations are generally treated as the relocation in position of geometric shapes, while at the secondary level and beyond, transformations deal more with the location of the graphs of functions and algebraic relationships. While the study of the former might theoretically be intended to inform the latter, in our experience the connection is too rarely made. Nevertheless, developing spatial reasoning and an understanding of symmetry are essential to studying and understanding transformations. Transformations also form the basis of many toys and games, and are used in various types of art, design, and decorating. This will be further explored in the last section of this chapter.

Exploration/Task

Choose any relatively flat object, for example, an irregular scrap of cardboard. Trace its outline on paper. Now move the object around in any way you please, including flipping it over, and retrace it. Compare the two outlines. How can you describe the movement and change required to move the object from the first position to the second?

Follow-Up and Discussion

In the exploration above, the second image was a *transformation* of the first. Producing the second image might have required any combination of transformations—and it may even be possible to produce the second image using a different set of these transformations.

A given mathematical shape or object on a plane (flat surface) can be *translated* (slid along the surface in any direction—sometimes called a *slide*), *reflected* (flipped over or reflected along a given line, so that what we see is the mirror image of the original shape—sometimes called a *flip*), or *rotated* (moved in a circular path around a centre point or centre of rotation—think of a sticker placed in the center of a rotating CD or DVD—sometimes called a *turn*). These movements can also be done in combination, and sometimes several different combinations of transformations will lead to the same end position.

Exploration/Task

Investigate each of the following transformations. See if you can find more than one way to get from each starting point to each resultant location (shaded). Mark any centres of rotation or mirror lines as required to illustrate the transformation.

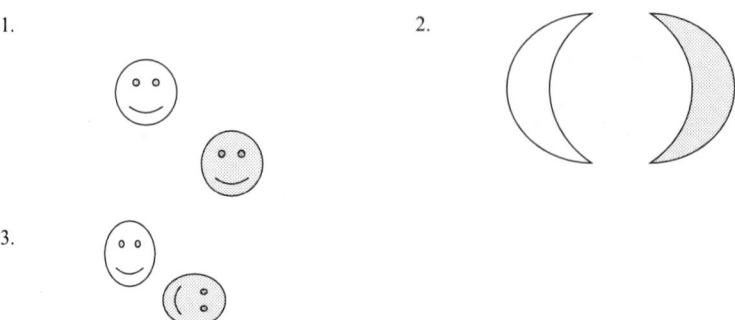

Follow-Up and Discussion

Many combinations of transformations could have been used in this task. For example, in 2, the resultant shape could be obtained by reflecting in the mirror line

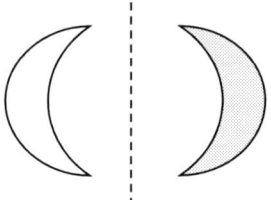

or by rotating around the centre point (among other possibilities).

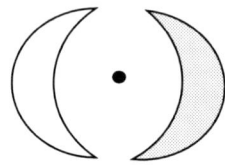

An important property related to transformations is *symmetry*.

Reflective symmetry means that a line can be drawn through a shape that acts like a mirror; for example the moon shape to follow has vertical reflective symmetry:

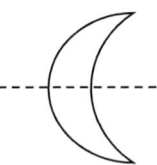

MATHEMATICAL TERM
Symmetry is a geometric property of being balanced around a line or a point (or a plane in three dimensions).

If we reflected the moon shape in the mirror line, we would not be able to tell the difference from the original. The mirror line drawn across the moon shape is a *line of symmetry*, and the shape has reflective symmetry in the line.

The following shape has *rotational symmetry* (as well as vertical and horizontal symmetry); we can draw lines of symmetry as shown in the illustration. If we rotated the shape in increments of one-eighth of a turn, it would appear the same.

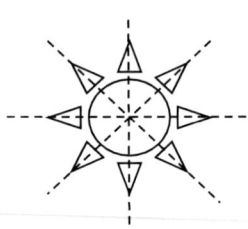

Practice and Further Exploration

For each shape below, draw any lines or points of symmetry you can see. Describe at least one transformation that leaves the initial shape unchanged and in the same location due to shape symmetry. In other words, describe a transformation that does not change the appearance of the shape at all.

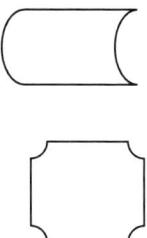

A *tessellation* is a shape that can fully cover a flat surface without gaps. Legend tells of a slave who was tiling a surface and broke some of the tiles; in order to avoid punishment, the slave came up with the idea of interlocking broken or irregular pieces. If we start with the following shape,

we can cut sections out of two of the sides and attach them on opposite sides as follows:

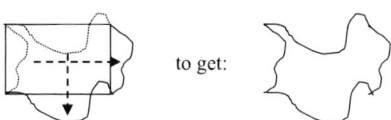

The resultant shape is a tiling pattern—you could tile a surface with it.

An important characteristic of the initial shape that will be used to create a tiling pattern in this way is that it must already be a tiling pattern. That is, before we transform the initial tile, it must itself be a tessellation tile. For example, a square or rectangle will completely cover (or tile) a flat surface, while a star or heart shape will not.

> **MATHEMATICAL TERM**
>
> A *tessellation* is a shape that can completely cover a flat surface with no gaps or overlap.

Practice and Further Exploration

1. Create a tessellation shape of your own using paper, scissors, and tape. Be sure your initial shape is itself a tile. (While hexagonal tiles will tessellate, doing so is more complicated, so we recommend initially using rectangular shapes). After cutting and reattaching parts of the shape as shown above,

trace it several times to show how it can be used to tile a surface. If desired, get creative and colour your pattern.
2. The artist M. C. Escher was famous for using mathematical tessellations in his art. Do a quick Web search of his name to view some of his work. Look for the tessellations. Students tend to particularly enjoy exploring transformational geometry using Escher's work.

13.7 Geometric Reasoning and Proof

The role of geometry in the curriculum has reached a turning point. Historically, traditional geometry on the plane (flat surface), named Euclidean geometry after the mathematician who was famous for exploring it centuries ago, was taught by having students focus on formal proof. Teachers introduced topics by stating definitions and axioms (generally accepted mathematical truths), then theorems were proved, and then students completed a series of exercises (Beaugris, 2013). Students were asked to construct diagrams using only a compass and a straight edge, and expected to follow (or often merely reproduce) formal mathematical arguments (proofs), all of which was thought to teach them mathematical reasoning.

For some time it has been argued that reproducing a rigorous mathematical argument when not developmentally ready to construct it oneself does not develop understanding or even rigor (Skemp, 1986), and that students are better off developing mathematical concepts using the kind of progression followed by mathematicians themselves when exploring and proving new ideas (Beaugris, 2013). Approaching proof by memorizing may in fact develop the sense that mathematics is about procedures and rules to remember (Handal, 2003). The reality in the past was, indeed, that many students simply memorized provided methods and proofs in order to get by (Holm and Kajander, 2011, 2012).

This traditional paradigm has evolved into more modern approaches, such as those described by NCTM (2000). As discussed in chapter 2, the work of Zack and Reid (2003) reminds us that reasoning and proving in the classroom must contain elements of *explaining* ideas, as well as verifying them, and the reasoning behind the arguments is of key importance (Harel and Lesh, 2003). Proof is a process of coming to understand, rather than a topic to be taught and learned (Reid, 2011, p. 15); as part of this process, learners should be taught to "think mathematically" (Beaugris, 2013, p. 31).

The Van Hieles were well-known researchers who studied children's developmental levels of geometric understanding and how they emerge. They suggested that students need to reach a certain level of mathematical development before formal proofs have meaning (Gutierrez, Jaime, and Fortuny, 1991). Today it is thought that true rigor emerges gradually, based on the development of sound mathematical *reasoning* (Stylianides and Ball, 2008), which can only be developed through deep mathematical experiences such as those described by Zack and Reid (2003, 2004).

More recently, developed curricula (for example, Mathematics 1–8, Ontario Ministry of Education, 2005) treat geometric reasoning as a process, approached in line with students' development. Rather than being asked to reproduce formal proofs, students are encouraged to explore the concepts, often concretely, and reason about them. In summary, the mathematical process of reasoning is a key component of the development of more and more rigorous arguments, which ultimately evolve into formal proofs when students are ready for them.

Practice and Further Exploration

Cuisenaire rods are one type of mathematical manipulative. They are plastic rods with lengths of 1 centimetre, 2 centimetres, and so on, up to 10 centimetres. If you don't have these plastic rods, you can construct a few similar rods by cutting out paper strips. They should be about 1 centimetre wide, and you will need one of each length from 1 to 5 centimetres. Use your strips to construct a mathematical argument about whether there is a strip in your set that is exactly half of the 5-centimetre strip. Think about the reasoning you used, and compare it to how a student might reason.

13.8 Technology and Dynamic Proof

Perhaps because of its historical roots, the use of traditional tools including straight edges, compasses, and protractors has been entrenched in the teaching of geometry. Although such tools seem to be going the way of the slide rule, they are still often cited in curricula. However, virtually none of the industries that use geometry—architecture, design, engineering, and so on—function without technology. The plethora of digital software packages and websites that support geometric constructions and explorations can greatly enhance learning.

Of particular use is software that has been specifically developed to support student explorations in geometry. Because such software allows for geometric constructions, as well as transformations of the geometric objects, the term *dynamic geometry software* is used to describe it. By constructing a generic example, such as a triangle that can be transformed or animated to represent *any* triangle, and measuring the sum of its angles, students can argue dynamically. This type of general example allows students to explore, reason, and argue about geometric properties. They will ultimately be able to provide reasoning and evidence that a given property—for example, that the sum of the interior angles of a triangle is 180 degrees—will always hold true. Such arguments are being called *dynamic proofs*, and are gaining greater and greater acceptance, even in the formal mathematics community. These dynamic arguments develop reasoning and deep understanding, and set the stage for the subsequent development of meaningful rigor.

Practice and Further Exploration

Explore some readily available electronic tools. We suggest the following:
1. The Illuminations activities on the National Council of Teachers of Mathematics site (found at http://illuminations.nctm.org/ActivitySearch.aspx).
2. The dynamic geometry software package Geometer's Sketchpad. Many school boards have access to it for free, and many tutorials are available on the Web.

3. The applets available on the National Library of Virtual Manipulatives website (found at www.nlvm.org).

13.9 The Pythagorean Relationship

Although students were traditionally taught the Pythagorean theorem as an algebraic formula, the relationship is in fact based on a geometric construct. Although the following exploration investigates the relationship using a concrete example, we also recommend using dynamic software if possible to construct a more general argument (as described above). Dynamic software allows students to construct generic examples—examples that can be seen to represent a *general* case—and, as a result, helps students get on the road to rigor more actively.

The idea of the Pythagorean relationship is that three lengths that exactly form the sides of a right-angled triangle (a triangle with one 90-degree angle) are then each used as the side-length of a square. We must ask: Is there a relationship among the areas of these three squares?

Exploration/Task

Having students use grid paper to cut out three squares with side-lengths that exactly form a right-angled triangle is a popular activity. While all triangle side-lengths that form a *right-angled* triangle can be used to illustrate the relationship, not all such triangles have sides with whole-numbered lengths. Triangles with sides that do not have whole-numbered lengths are more difficult to model on grid paper, and thus the relationship is more difficult to verify by calculation. The easiest lengths to use are 3, 4, and 5, but side lengths of 6, 8, and 10, or 5, 12, and 13 can also be used. Follow these steps:

1. Construct (using grid paper) and cut out three squares with side-lengths corresponding to any one of the whole-numbered sets provided above.
2. Arrange the squares to show that bringing together one side of each will form a right-angled triangle.
3. Place one of the smaller squares on top of the biggest one.
4. Cut out the other smaller square, and fit it into the remaining area.

How do the areas compare?

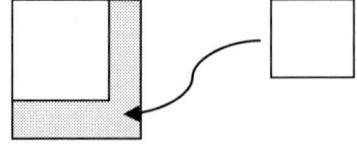

Verify your observation by using the following steps to calculate the areas numerically:

1. Calculate the area of the largest square (using length times length, or by simply counting the unit squares inside).
2. Calculate the area of the two smaller squares, and add these two areas together.
3. Compare the larger area with the *sum* of the two smaller areas.

The sum of the two smaller squares' areas should equal the area of the larger square. Note also that the largest square was the one constructed using the length of the side of the triangle opposite to the right angle.

Follow-Up and Discussion

In the above task, you should have found that the areas of the two smaller squares exactly cover the larger one; the sum of the area of the two small squares is literally the area of the large square. Mathematicians would argue that a single example is insufficient to form a general conclusion. What if you redid this activity 30 times, with students choosing, measuring, and cutting random right-triangle-length squares? You might have stronger evidence that this theory is true, but it is still not a proof. However, if you were able to construct a generic example using software—to show that the relationship was true no matter what size right-angled triangle was used—you would have even stronger evidence. Such dynamic proofs are a middle ground between using examples and a more formal proof. As students move from a specific example to a generic example, they are developing reasoning. Moving on to a dynamic proof supports a more general and rigorous argument. Such stages are important in the development of reasoning and more formal proof.

Indeed, many more formal proofs exist to show that the Pythagorean relationship is always true. It turns out that no matter what the size of the right-angled triangle, the two smaller squares always have a combined area equal to the area of the larger square. The largest square is the square formed using the length of the diagonal (the side of the triangle opposite the 90-degree angle), and this side is called the *hypotenuse*. For example, in the 3, 4, 5 case, the area of each square is calculated as follows:

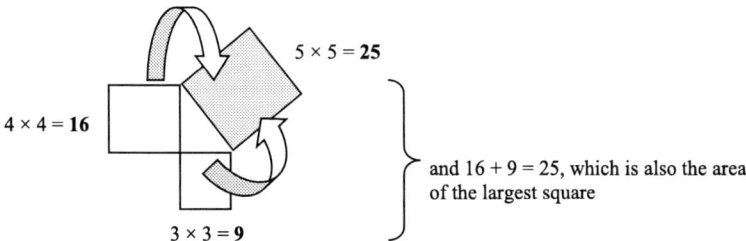

and $16 + 9 = 25$, which is also the area of the largest square

Thus the Pythagorean relationship is really about right-angled triangles and the squares constructed on each of the triangle's three sides; however, it is often stated more algebraically. If a, b, and c are used to represent the lengths of the triangle's three sides, with c being the hypotenuse or longest side, then the area relationship is:

$$(a \times a) + (b \times b) = (c \times c)$$

or

$$a^2 + b^2 = c^2$$

In fact, this relationship can be handy as an algebraic tool, for situations in which we know two measurements along sides of a right-angled triangle, but not the third. For example, if we wanted to know a diagonal distance, but had only the two perpendicular measurements, the relationship could be used.

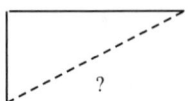

Practice and Further Exploration

The GeoGebra application (found at http://www.geogebra.org/en/upload/files/english/Flint_Jessica/PythagTheorem.html) allows students to change the triangle size, and see the algebraic and geometric relationships together. Spend a little time exploring this activity.

13.10 Geometry in Art, Design, and Entertainment

Geometry has great potential to appeal to students' aesthetic senses, as many functional aspects of art—such as perspective, vanishing points, and the golden ratio—have mathematical origins. Connecting geometry to art whenever possible enriches both subjects. Students can construct tessellations in art class and discuss them later in math class, or investigate patterns and designs using both mathematical and artistic lenses.

Fractal geometry is a new area of geometric study, discovered quite recently, which is based on patterns that are repeated (or *iterated*) over and over. Mathematicians have been surprised to learn that some of these abstract geometric shapes do an excellent job of modelling nature, such as root systems, blood vessels, and some geographic phenomena. Students may be fascinated to explore some of these images and applications on the Web.

Geometry also plays a key role in design and animation, and many toys and amusements have a geometrical basis. For example, the traditional kaleidoscope makes use of a triangular prism of mirrors facing inwards to construct an image. Kaleidoscopic images are wonderful examples of transformational geometry at work. The result of the triangular-prism-shaped mirrors is an image that looks like a 60-degree-wedge shape reflected six times in each of six radii around a central point:

Practice and Further Exploration

1. Explore fractal geometry on the Web. Choose and print out one of your favourite images, and explore the patterns within.

2. Construct a kaleidoscope pattern on paper using the ideas of transformational geometry, as described in the steps to follow. You will need paper, scissors, a black marker, and a protractor or a set of fraction circle manipulatives.

   - Trace a one-sixth piece from a set of fraction-circle manipulatives or measure a 60-degree angle with a protractor and cut a wedge of that angle out of a paper circle.
   - Use the black marker to draw an asymmetrical design on the paper wedge, going right to the edges of the paper wedge.

   - Turn the paper wedge over, and copy the image onto the *back* of the paper wedge by holding it up to the light so you can see through the paper.
   - If circle manipulatives are available, trace the one-whole fraction-circle piece on paper, or draw a circle the same size as the one that the original wedge was cut from.
   - Place the wedge pattern template *under* the paper circle, lining up the point of the wedge with the centre of the circle.
   - Trace the pattern on the wedge. For a more kaleidoscopic effect, do *not* trace the *edges* (radii) of the pattern wedge—the dotted lines shown below—onto your design. You may need to hold the paper up against a window to see through to the pattern beneath.

   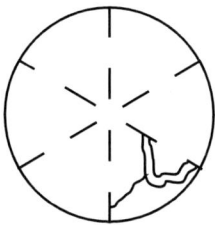

   - Next, model the reflection created by the mirror in a kaleidoscope. *Flip* your wedge pattern template over and place under the next blank wedge in the circle. Trace it again, producing the mirror image of the first traced wedge.

   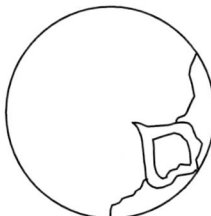

3. Continue flipping and tracing the pattern in each wedge around the circle. There should be six reflected images in all. You may colour in your kaleidoscope pattern if you like.

Chapter Problems

1. Draw a 3-centimeter-by-5-centimetre rectangle, using a 5-centimetre length as the base. Beside it, draw two different non-rectangular parallelograms, each with a base of 5 centimetres and a height of 3 centimetres. Label them parallelogram A and B, respectively.
   a) Compare the areas of all three shapes. Justify your answer.
   b) Compare the perimeters of all three shapes. Justify your answer.
   c) What properties can you suggest about non-rectangular parallelograms in general, when compared to the rectangle with the same base and height?
2. Find a cereal or cracker box. Measure its dimensions in centimetres. Determine how many solid plastic centicubes would fit in the box. Assume that you cannot cut the plastic centicubes into smaller pieces. Think about how students could relate the number of cubes to the volume.
3. Draw a square. Inside the square, draw a circle that touches each side of the square at a single point.
   a) Inside the circle, draw another square with each vertex intersecting the circle circumference, and with its base parallel to the first square. Find the ratio of the area of the two squares to each other.
   b) Draw the picture again, but this time, when you draw the inner square, rotate it 45 degrees so that it looks like a "diamond" shape. Find the ratio using a different method than in a).

Further Reading

Benson, C. C., and Malm, C. G. (2011). Bring the Pythagorean theorem full circle. *Mathematics Teaching in the Middle School, 16*(6), 336–344.

   Benson and Malm discuss visualizing the Pythagorean theorem.

Cox, D. C., and Edwards, T. (2012). Sizing up the Grinch's heart. *Mathematics Teaching in the Middle School, 18*(4), 228–235.

   Cox and Edwards use the context of the children's story *How the Grinch Stole Christmas* to create activities for exploring concepts of similarity in shapes.

Kestell, M. L., and Kubota-Zarivnij, K. (eds.) (2012). *Abacus, 50*(3).

   This entire issue of *Abacus* is dedicated to geometric transformations. It includes a discussion of the mathematics of some of the concepts, as well as common errors and misconceptions in classrooms. The issue includes manipulatives, as well as a classroom lesson plan that can be used to teach geometric transformations to elementary students.

Neel-Romine, L. A., Paul, S., and Shafer, K. G. (2012). Get to know a circle. *Mathematics Teaching in the Middle School, 18*(4), 222–227.

   Neel-Romine, Paul, and Shafer detail a hands-on activity that uses Play-Doh to help students explore their definitions of a circle, and create a more robust definition for the two-dimensional shape as a result of the exploration.

Nivens, R. A., Peters, T. C., and Nivens, J. (2012). Views of isometric geometry. *Teaching Children Mathematics, 18*(6), 346–353.

The authors discuss using isometric dot paper to draw cubes and how to connect the activity to real-life applications.

Zack, V., & Reid, D. A. (2004). Good-enough understanding: Theorising about the learning of complex ideas (part 2). *For the Learning of Mathematics, 24*(1), 24–28. http://www.acadiau.ca/~dreid/publications/ZackReid2_2004flm.pdf

The authors show examples of students' reasoning as it grows and deepens in geometric contexts.

Chapter 14
Measurement

14.1 Linear Measure

Children often begin to understand measurement by estimating lengths both informally and using a standard unit. These experiences are critically important; indeed, if students have difficulties with measurement in later grades, this may be the result of insufficient experience actually *measuring* things.

The two systems of measurement in use in North America continue to cause some confusion for students (and some adults!). In Canada, children are taught the metric system in school, yet retailers continue to offer goods using imperial measures, such as lengths sold in feet or yards, as is done in the United States. From a practical standpoint, some familiarity with both systems is helpful.

From a mathematical and scientific point of view, the metric system has some beauty. Deeply dependent on the ideas of the decimal number system, the concepts and units in metric measure directly align with concepts of (base-ten, whole-number) numeracy, place value, and operations, as well as decimals. The central relationship of relative positions and values in the decimal system—ten times as much or a tenth as much as we move left or right—align with units of linear measure.

Beginning with the standard linear measure of one metre, we can either subdivide the one-metre unit or multiply it by factors of ten. (Note: It is helpful for children to be able to see and touch a metre stick while talking about it.) Again, purposeful connections to ideas of place value—and visualizing by *looking* at an example of the size of the unit when possible—can be helpful. Here are some common linear units:

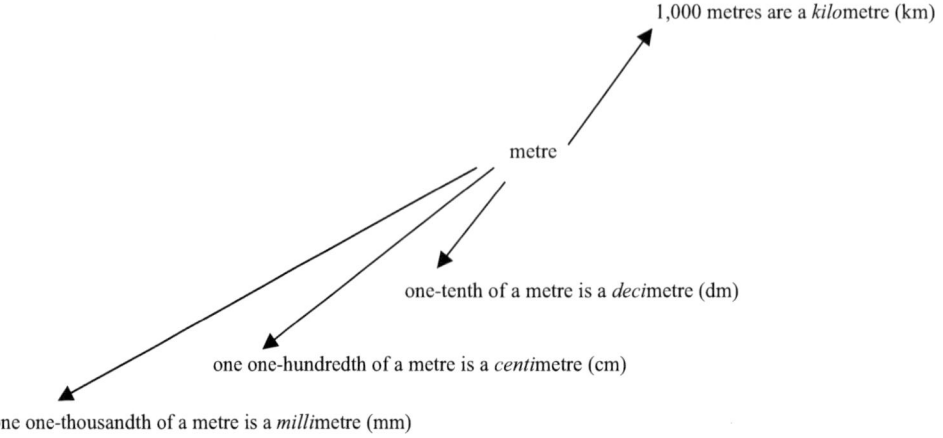

The prefixes—milli, deci, and so on—are important parts of the unit. While at some level these do need to be remembered—and word walls can help students do so—looking at the words themselves can also help. For example, *centi* has a connection to the French *cent* for one hundred, but also to the idea of Canadian and American money, where one cent is one one-hundredth of one dollar. *Deci* reminds us of the decimal (base *ten*) system. Mastering these meanings now will help students later when they are learning other types of units for area, mass, and so on.

There is a certain tendency to focus the study of measurement in school on conversions from one unit to another. As we suggested earlier, actually *measuring* things and working on contexts in a hands-on manner may better support student understanding of both measurements and conversions. We also suggest teaching students to perform conversions, not by using memorized algorithms, but by reasoning, while referring to the units on a manipulative such as a metre stick.

A typical procedural error made by a student who is asked to convert 80 centimetres to metres is to think, *there are 100 centimetres in a metre, so the answer must be 80 × 100*. That would be 8,000 metres, or 8 kilometres. A quick look at a metre stick makes it clear that 80 centimetres is less than even one metre. How does such mistaken reasoning come about when performing a conversion, and what are the misconceptions that have led to this error? How can we provide an environment in which such misconceptions do not so easily occur?

Exploration/Task

Explore the problem of converting 80 centimetres to metres by constructing a suitable model. Convert the length correctly, and think about how you might address the student's reasoning described earlier.

Follow-Up and Discussion

By rethinking the *how many metres in 80 centimetres* task, and physically looking at a representation of 80 centimetres, we see it is equivalent to *less* than a full metre, or 100 centimetres. The number of metres that is equivalent to 80 centimetres must be less than one. One way to think about it is that 80 centimetres is 80/100 of a metre, or 0.8 metres.

Students can be asked to explore other concrete examples, and think about the reasoning required to convert the given amounts. A scaffold might be to ask them to consider 50 centimetres. Looking at the metre stick, they can see that 50 centimetres is one-half of a metre—or 0.5 metres. Children may need to explore a number of examples before they can construct the generalization of this concept, which they might express as something like this: *If the unit is larger, we don't need as big a number part for the same amount. But if the unit is smaller, we need more of them.*

> **INSIGHT**
>
> When using a larger unit (for example, metres rather than centimetres), the number part will be smaller when representing the same length.

Practice and Further Exploration

Practice your visualization skills by completing the following conversions. Be sure to use a metre stick or other model. Think through the reasoning that a student might use as you work through each question.

1. 5 m to cm
2. 20 mm to cm
3. 15 dm to m
4. 0.6 km to m
5. 5,670 cm to km (You might want to double-check this one by converting it back to centimetres again.)

In the following section, we will look at using such units to measure objects and their attributes.

Perimeter

Children can best develop the intuitive notion of perimeter, which is the boundary of a shape, by moving their bodies along the edge of various shapes. For example, circles are painted on the floor of many gymnasiums. Alternatively, shapes can be drawn on pavement in chalk. Walking around these shapes gives students a physical sense of perimeter. We can ask, "How far did we go in total?" or "How far did we walk to get back to where we started?" Such activities emphasize the measurement aspect of perimeter—it is a distance. We already know that distances are measured in linear units, such as feet or metres.

Some confusion may arise for students in relation to the term *linear*. In everyday language, *linear* suggests a straight line. But the shape students walk around on the schoolyard may not have only straight sides. The term *linear*, in mathematics measurement, is not quite literal. Rather than implying a straight line or distance, it implies the idea of something that *can be measured* by a straight-line unit. For example, imagine carefully placing a string around a large (possibly curved) shape on the playground and cutting the string off at exactly the right length to meet the starting point. We could then lay the string out in a straight line, and measure it with a linear unit.

Some geometric shapes have properties that make them easy to measure. For example, a square has four sides of the same length, so knowing the length of just one side is enough to find the perimeter. For shapes with different side-lengths, it may be necessary to measure each side.

Although it is tempting to provide a formula for students to use to calculate the perimeter of a common shape such as a rectangle, we recommend letting students add up the lengths of the sides at first. Later they will realize that they can also measure the two sides of different lengths, and double each.

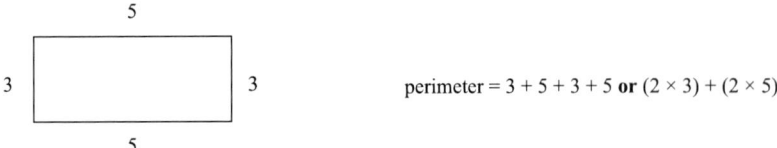

Curved shapes, however, are more challenging to measure. An easy way to measure the perimeter of many curved shapes is by placing a string around the shape to duplicate the distance of the perimeter, and then laying it out and measure it. The word *circumference* is the name for the perimeter of a circle. We could get by without knowing this word, and could simply refer to the perimeter of the circle. Because the circle is a frequently used shape, and is also mathematically interesting, we will investigate its perimeter in more detail.

The following task can be used as an introduction to help students *estimate* the circle perimeter or circumference. Later in the section, we will derive the exact rule or formula for calculating this measurement.

MATHEMATICAL TERM

Circumference is the term given to the perimeter of a circle.

Exploration/Task

Using pennies (or round counters or other circular manipulatives), create rings of pennies around one central penny. Marking the centre penny with a sticker might help you keep track of the centre throughout this process.

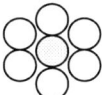

Follow-Up and Discussion

We can see that the first ring contains 6 pennies. You should have found that the second ring contains 12, and, if you kept going, that the next ring contains 18.

Reasoning

Each time we add a ring, we effectively add 2 pennies to the complete shape *diameter* (any straight line that passes through the centre of the circle, with end points on the circumference), and 6 to the number in the outside ring. We notice an important pattern as we create each new ring: adding 2 to the diameter adds 6 to the circumference. Alternatively, we could look only at the *radius* (the distance from the centre of the circle to any point on its circumference) and see that adding 1 penny to the radius adds 6 to the circumference.

These observations might suggest that when the diameter gets bigger, the circumference gets bigger by about 3 times that much, or when the radius gets bigger, the circumference gets bigger by about 6 times as much. Trying this again using a different size of circular manipulative, such as larger coloured counters, provides further evidence that this property holds generally.

The observation that the circumference grows by about 3 units for every new unit of diameter, is a perfectly good working estimate of the relationship between the circumference and the diameter of the circle. In fact, "tree huggers" use this definition to determine the diameter of trees in the forest. If you hug a tree to estimate the circumference of the (circular) trunk, you can estimate its diameter before cutting it down, since the circumference (C) is about 3 times the diameter (D), or the diameter is approximately the circumference divided by 3. We conjecture that C is about 3 × D, and D is about C ÷ 3.

Even more accurate versions of the circle circumference/diameter relationship can be found through accessible classroom exploration. String can be used to measure the diameter and circumference of several large circles on the gym floor or drawn in the schoolyard. (Note: The circles must be drawn accurately—use a piece of string tied to a piece of chalk to carefully draw the circumference.)

> **MATHEMATICAL TERM**
>
> The *diameter* of a circle is any straight line that passes through the centre of the circle, with end points on the circumference.

> **MATHEMATICAL TERM**
>
> The *radius* is the distance from the centre of a circle to any point on its circumference.

Exploration/Task

Several examples of student-generated diameter and circumference measurements are listed in the following table; however, students can and should be asked to gather their own measurements. What relationship might you find between circumference and diameter using the measurements? (Use a calculator for accuracy.)

Diameter	Circumference
80 cm	2.6 m
1.5 m	4.7 m
3.2 m	9.8 m

Follow-Up and Discussion

Looking at the table, we might see that, as expected, the circumference is about 3 times the diameter. One way to test this is to multiply each diameter measure by 3; for example, for the first value in the list, 80 cm × 3 is 240 cm or 2.4 m. If the circumference measurement is divided by the diameter in each example, the values are 3.25, 3.13, and 3.06. If we averaged these three values (that is, added them up and divided the sum by 3), we would get 3.147. This estimated value, 3.147, is our conjectured ratio value based on our given data—if we multiply another diameter by this value, we would get an approximation of the circumference.

If we calculate the C/D ratio for a list of measured values, the more values we take, the closer the average of these values often is to 3.1 or even 3.14. Students can actually discover the strange and exciting number 3.14159... that goes on forever, which we call *pi*. Pi is the pronunciation of the Greek letter π that is used to represent this (infinitely long) value. Since we can't write down all of its digits, we have to represent it somehow, and the Greek letter π is historically how this was done—and still is.

The number π has fascinated mathematicians for centuries. It is becoming popular to celebrate this number on—you guessed it—March 14 or 3/14. Some classrooms have a fun mathematics day typically called Pi Day, full of circle-related activities, on March 14. A quick Web search will locate many classroom-ready Pi Day activities. (And, by the way, March 14 is also Einstein's birthday.)

If you have access to dynamic geometry software (described in section 13.8), students can measure the circumference and diameter of circles more accurately using the measurement functions. Depending on how many decimal places the software has been set to use, children can calculate π to any number of digits. It is important to know that all of the values 3, 3.1, and 3.14 are *approximations* of π. We can use as many digits as we need, based on the required accuracy, but none will be perfectly accurate.

Expressed in formal language, the perimeter formulae for circles are:

$$C = \pi \times D$$

Or, knowing that D = 2 × r,

$$C = \pi \times 2 \times r \text{ or } C = 2 \times \pi \times r$$

> **INSIGHT**
> The digits of the number called π (or *pi*) can never all be written down in decimal form because it is infinitely long.

> **INSIGHT**
> The perimeter of a circle is about 3 times its diameter. To be more accurate, the value of π to any number of digits can be used instead of the 3.

14.2 Rectangular Area: Exploring Covering and Counting

Rectangular area refers to area that can be covered with unit tiles, such as the two-dimensional space on the floor of a rectangular room. Unit tiles are tiles that measure one unit by one unit; they are a handy way of determining the size of this space. The idea of measuring area is an example of a concept that becomes quite easy if students have learned previous concepts about multiplication models conceptually. If the product of two numbers was already explored using a rectangular area model (see section 5.2), then measuring area that can be drawn as a rectangle can be directly connected to this idea.

Area of a Rectangle

Exploration/Task

Consider the following 3-by-5 space, which we can cover with 1-unit-by-1-unit squares (called unit squares) as shown below. Find a way to count the number of squares.

Follow-Up and Discussion

Students can be encouraged to count the squares as a way of finding the area they represent or cover. Students who are familiar with the area model for multiplication will often immediately see that the number of square tiles is equal to the dimensions of the rectangle multiplied together: 3 × 5 or 15. This observation then generalizes to the length-times-width formula. A more general way to express this formula is by saying base times height. Using the word *base* removes the problem of which is the length and which is the width. For example, the base of the rectangle shown is 5 units and the height is 3 units.

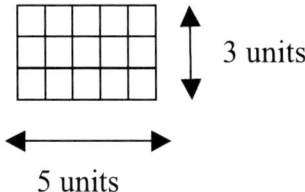

We call one of the small squares in the rectangle a unit square. But what if the side-lengths were measured in centimetres? Or metres? The units for area measurement are different from those used to measure length.

Our rectangle is drawn with sides of 3 lengths of 1 centimetre and 5 lengths of 1 centimetre. Multiplying 3 by 5 gives us 15. The model shows 15 of the 1-centimetre-by-1-centimetre squares. Numerically, we can think:

3 cm × 5 cm, or
3 × 5 × cm × cm, or
15 × cm × cm, which is sometimes written
15 cm², which we read as, "15 *square* centimetres."

It is important to remember that cm² refers to a fixed unit rather than an operation. At first, it may be simpler to write the unit out in words, such as *square cm*. It is important that children can visually see and touch square centimetres, so that they can recognize that they are distinctly different from lengths, as illustrated below:

☐ *1 cm² is like a little square with each side having length 1 cm*

Using Rectangles to Find Areas of Other Shapes

> **INSIGHT**
> If we can rearrange an area to form a rectangle, and if we know the base and height of the resulting rectangle, we can find the area of a given shape.

> **CONNECTION**
> See chapter 13 to review the geometry of these shapes.

In the previous discussion, we explored how it makes sense that the area of a rectangular shape can be found if we know the base and the height. We can now use our understanding of the area of a rectangle to determine the area of many other shapes, as long as they can be *rearranged into a rectangular shape*.

In the next few paragraphs, we will explore how we can rearrange the area of several common two-dimensional shapes—a parallelogram, a triangle, and a trapezoid—into a rectangle.

Once we rearrange these shapes, their areas can be found using what we know about the area of a rectangle. Students can conduct such explorations by cutting shapes out of paper, but it is even better if they use dynamic software (see section 13.8). The advantage of dynamic software is that it makes it easier for students to see that a given construction will *always* work.

Exploration/Task

For this activity, you will need drawings or cut-outs of the following three shapes. Rearrange the area of each in a way that allows you to use what you know about rectangles to determine the area. The small arrow symbols on pairs of sides indicate that the sides are parallel.

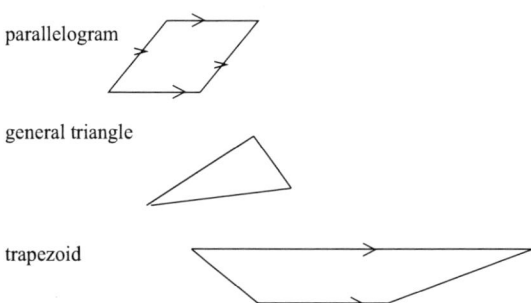

Follow-Up and Discussion

In each case, the key idea is finding a way rearrange the area of the original shape so that you can see it as a rectangle. The following constructions illustrate this concept:

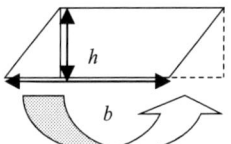

Rearranging part of the parallelogram area shows that the parallelogram has the same area as a rectangle with the same base and height. The formula for area is exactly the same as a rectangle: $b \times h$.

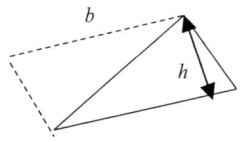
By duplicating the triangle and rearranging it, we can make a parallelogram. Further examples might be needed to determine if this always works. The height (h) must still be perpendicular (at a 90-degree angle) to the side used as the base (b). The triangle is half of the parallelogram, and we already know the parallelogram area from the previous diagram. Half of the parallelogram with the same base and height is $1/2 \times (b \times h)$. This is also sometimes written $(b \times h)/2$.

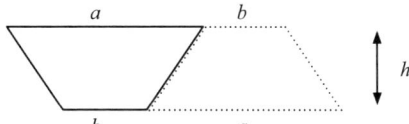

There are a number of ways to visualize the area of a trapezoid. One way is to duplicate the shape to create a parallelogram. The new parallelogram shape has the same height as the trapezoid, but a base that is twice as wide. The newly created parallelogram as shown has area base times height $= (a + b) \times h$, but this encompasses *both* trapezoids—it is double what we need. Consequently, for the trapezoid formula we simply divide by 2: $(a + b) \times h/2$. There are other ways to find this trapezoid area formula as well. The following diagram suggests one way:

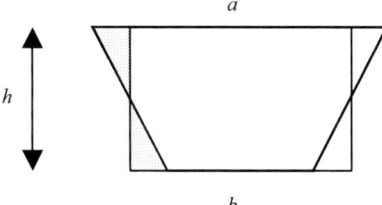

This construction shows that we can construct a rectangle with the same height as the trapezoid, but with a base that is halfway between the original top and bottom lengths of the trapezoid, or $(a + b)/2$. In fact, this new length, the base of the newly constructed rectangle, is the *average* or midpoint length of a and b. Multiplying this expression for the rectangle base $(a + b)/2$ by the height gives the same formula for calculating the trapezoid area, $((a + b) \times h)/2$, as we discovered previously.

Units and Conversions

Converting areas from one unit to another presents several conceptual challenges. Let's start by considering the following context.

Example

Chocolates can be ordered individually wrapped, as well as in bulk boxes of 100. The boxes have 10 chocolates in each of 10 rows:

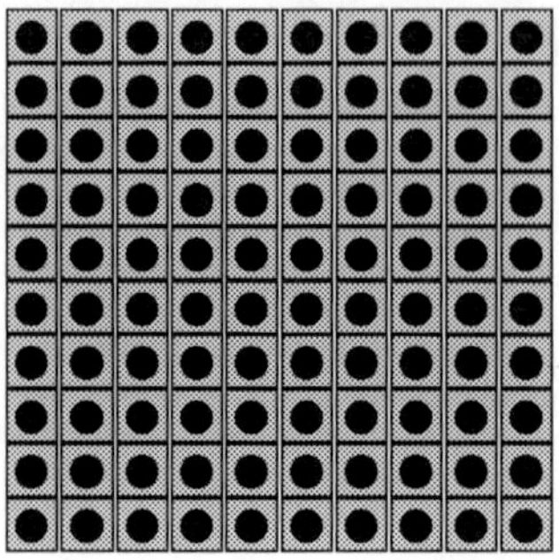

If we want 400 individual chocolates, how many boxes would we need? By looking at the above illustration, it seems clear that we need 4 such boxes to get 400 chocolates (since 4 × 100 is 400).

Exploration/Task

1. Answer the following decontextualized problem using a method of your choice: Convert 400 square centimetres to square decimetres. Remember that there are 10 centimetres in 1 decimetre, but remember also that you are working with square units.
2. After answering the question, consider the following possible student answers. Think about the reasoning they might have been used to determine these answers. Which of these is correct?
   400 cm² can be expressed as:

   - 40,000 dm²
   - 40 dm²
   - 4 dm²

Follow-Up and Discussion

Without a picture or context, the task problem seems quite difficult. This problem is *not* a good example of a classroom problem. It does not have a context, offer a purpose for solving it, or prompt students to use a model to find the solution.

The problem could be solved using a diagram or context similar to the box of chocolates. A square decimetre is a 10-centimetre-by-10-centimetre square—similar to the drawing of the box of 100 chocolates. In fact, the problem is exactly analogous to the chocolates problem in the example, and the answer is 4 decimetres². What might the students who answered 40,000 or 40 decimetres² have been thinking?

To get an answer of 40,000, the student likely multiplied by 100 to get from centimetres² to decimetres². This is a common misapplication of the reasoning that "1 decimetre² is 100 times as big as a centimetre², so to convert we multiply by 100." Looking at the *area* in 1 decimetre² (by checking the chocolates diagram), we see that using a bigger unit (the boxes of chocolates, or decimetres²) requires *less* of them. In the 400 chocolates case, when we moved to the bigger unit (the box) we needed *fewer* boxes than individual chocolates, rather than more. Similarly, when converting from centimetres² to decimetres², we need fewer decimetres² than centimetres². While this concept is not difficult to understand in the context of the boxes of chocolate, it is much more difficult in the abstract and without a model.

To get the answer of 40, it is possible that the previous misconception is present as well as one other: the student may have forgotten that a square decimetre is 10 times as big as a centimetre *in two directions*—100 times in all. The student may have multiplied by 10 using only the reasoning that there are 10 centimetres in a decimetre.

Practice and Further Exploration

Use reasoning, sketches, and models to solve the following questions:

1. Using the information about the chocolates provided above, namely that there are 100 chocolates in a box, express 20 chocolates as a portion of a full box of 100. Use fractions, decimals, or any other representation you choose. Compare this question to the following problem: *convert 20 centimetres^2 to square decimetres*.
2. A field measures 1,500 metres by 900 metres. Draw a model and express this area in square kilometres.
3. You are told that 5.6 metres^2 of fabric are available for a craft project, but you don't know the exact dimensions of the fabric. Find two different sets of possible rectangular dimensions for the fabric, expressed in centimetres. (Note: There are many different possible answers.)
4. Express 3,650,000 millimetres^2 in metres^2. Use a diagram to help you. It may also be helpful to translate into square centimetres first.

14.3 Circle Area

An important concept established in the previous section is the idea of rearranging areas to find an easier way to measure them. We will need to use this technique to determine the area of a circle. The circle doesn't seem to immediately lend itself to our previous technique of creating rectangles from other shapes, unless we are a bit creative. We will first explore a method for estimating circle area, followed by another slightly more challenging method that allows us to find the exact area.

Estimating Circle Area Using Squares

Exploration/Task

Draw a circle and two squares as shown below; one square will be inside the circle, and one outside. We will label the *radius* of the circle as r using a dotted line.

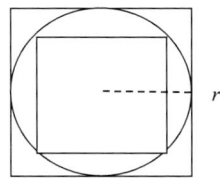

In the diagram, the circle area seems to be less than the area of the outer square, but greater than the area of the inner square. You might want to draw in some more circle radii to help you more easily see the dimensions of the outer square and the inner square, in terms of the circle radius r:

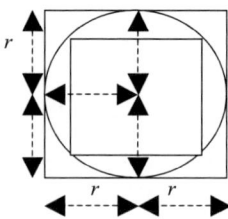

Labelling multiple occurrences of the original radius length r helps us see how to measure the area of both squares in terms of r. Using the lengths of r as marked, and others that you may identify, as well as what you know about square and triangular area measurement, find an expression in terms of r for each square's area.

Follow-Up and Discussion

Both of these squares can indeed be expressed in terms of r. Both sides of the outer square have length $(r + r)$ or $2 \times r$. Multiplying $(2 \times r) \times (2 \times r)$ gives the area of the big square as $2 \times r \times 2 \times r$ or $4 \times r^2$.

We can think of the small square as two triangles of base $2 \times r$ and height r. (To see this, we need to draw in even more lines to indicate the radius—those going from the centre to each vertex or corner of the inner square.) The area of each of these triangles is $\frac{1}{2} \times (b \times h)$ because a triangle is half of the corresponding parallelogram. In our case, the base is the diagonal or $2 \times r$, and the height is r, giving us $\frac{1}{2} \times (2 \times r \times r)$. The small square contains two of these triangles; therefore, $2 \times [\frac{1}{2} (2 \times r \times r)]$ is just $2 \times r \times r$, or $2 \times r^2$.

This investigation tells us that the circle area must be somewhere *between* $2r^2$ and $4r^2$, because the circle is inside the $4r^2$ square and outside the square with area $2r^2$; therefore, the area of the circle must be between these two values. Taking the middle value, we get an estimated formula of $3r^2$, which is a good estimate for the circle area.

Deriving the Formula

While the following method provides an accurate formula, it draws on some more intuitively deep mathematical concepts. Students are generally able to appreciate the reasoning involved, if approached as follows. We will again use the idea of subdividing and rearranging the area, but will also draw on the intuitive notion of mathematical infinity; students will need to keep doing the subdivision over and over again in their minds. To derive this formula for area, we also need to know the formula for the circumference of a circle from section 14.1: $C = 2 \times \pi \times r$.

Using a paper circle and scissors to follow along with the guided exploration below will help students understand the constructions involved. (We recommend that readers try this, too.)

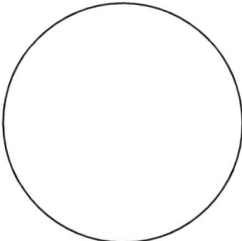

Divide your circle into eight pie-shaped wedges of equal size.

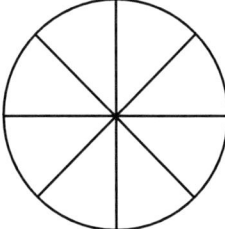

Use scissors to cut the wedges apart. Arrange them in a rough parallelogram as illustrated below:

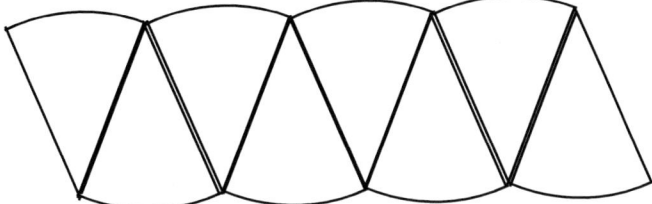

It is a bit of a stretch at this point to suggest that the shape formed by placing the wedges in a row resembles anything familiar. It certainly isn't a rectangle. But what if we *keep on cutting*? That is, what if we cut each wedge vertically (along the vertical height, or the vertical radius), into two wedges? And then keep rearranging the wedges into alternate positions as before?

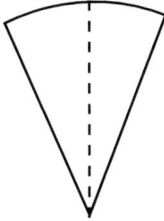

Try this at least twice more. After several iterations, the circle shape looks much *more* like a rectangle than it did on our first attempt.

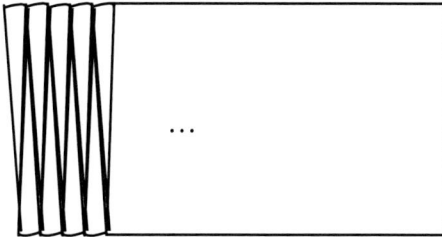

Reasoning

At this point, we need to draw on an important mathematical idea: the intuitive sense of moving closer and closer. If we had used a computer to subdivide the wedge, we could see that every time we did so, the illustrated area got *closer* to the area of a rectangle. Eventually, we would get a very accurate representation of a rectangle, one with an area that is as close as we want to the original circle. Realizing that we can get infinitely close to an exact rectangle allows us to generalize that the rectangle *is* a way to measure the circle area.

Next, we need to find the dimensions of such a rectangle made with infinitely subdivided wedges (our new representation of the original circle). The rectangle height comes directly from the circle's radius:

> **INSIGHT**
> Mathematics allows us to repeat a process—infinitely!

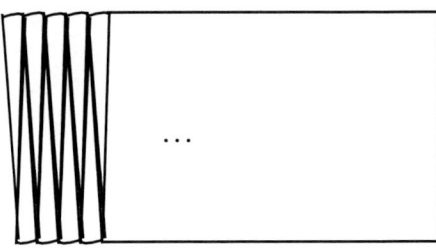

r The rectangle height is exactly the original circle's radius.

The rectangle base also comes from the original circle. Looking at the first shape formed using the eight wedges makes it clearer that the base is only *half* of the original circle circumference. The other half of the circumference forms the top length of the rectangle. Since the circle circumference is $2 \times \pi \times r$, half of it is just $\pi \times r$. The rectangle base is $\pi \times r$ or just πr. Examining our almost-rectangle, we can now write the dimensions:

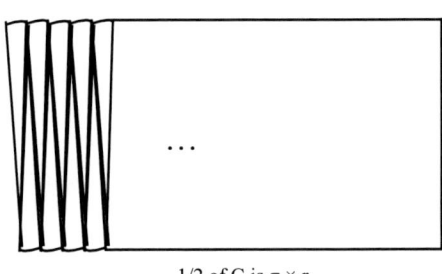

1/2 of C is π × r

r The rectangle height is exactly the original circle's radius.

Therefore, the area is p × r × r

Reasoning

By multiplying the base and height, we get the rectangular area—which is the circle area of $\pi \times r \times r$. We sometimes write a shortened version of the circle area formula as πr^2; however, many teachers prefer to use the $\pi \times r \times r$ version of the formula, because their students find it more accessible without the additional challenge of interpreting the exponent. Knowing that π is about 3 aligns well with the prediction made using the earlier estimation method, which was that the area would be between $2r^2$ and $4r^2$. We see πr^2 is indeed between $2r^2$ and $4r^2$.

14.4 Relationships between Area and Perimeter

One might initially think that as the perimeter of a shape gets larger, so does the area. The problem with this conception is that it is true sometimes, and untrue at others. Students can explore the perimeter-area relationship using toothpicks, for example, as a handy unit of length.

Exploration/Task

Using 12 toothpicks as a rectangle perimeter measure, make a number of different rectangles. Calculate the area of each rectangle as you work, using the established base-times-height relationship for area. For this task, we will define the new area unit to be *square toothpick units*.

Follow-Up and Discussion

It doesn't take long to find two rectangles with a perimeter of 12 toothpicks that have different areas:

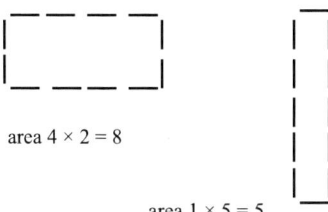

You might have also noticed that the square rectangle of perimeter 12 is in fact a 3-by-3 square, with an area of 9. Note that if you listed all the possible rectangles that could be formed using 12 toothpicks, the square is the largest in area. By investigating the various areas of rectangles with the same perimeter, we can make a conjecture about whether the square rectangle will always have the greatest area.

Exploration/Task

List all of the possible rectangles with a perimeter of 24 to find the one with the largest possible area. Make a prediction about the largest area of a rectangle with a perimeter of 32 and then test your prediction.

Follow-Up and Discussion

The largest rectangle with a 24-unit perimeter has an area of 36 square units (the 6-by-6 square) and the largest rectangle with a 32-unit perimeter has an area of 64 square units (an 8-by-8 square). In each situation, we observe that the largest rectangle is the square one.

Reasoning

At the elementary level, the use of multiple examples is usually taken as sufficient to argue that the square will always be the largest rectangle for a given perimeter. At the secondary level, this idea is constructed more formally. It is in fact possible to provide at least some further evidence for this claim without secondary-level mathematics, as will be shown below.

This generalized version of the question can be explored more easily using charts of numbers. For example, for a perimeter of 28, the sum of the base and height, $b + h$, must be half the perimeter, here 14. (Note that we could express all these values algebraically, but things would get much more confusing.) What is the biggest product possible of b and h for such rectangles? Provided are both a chart and graph of the areas for different base measurements.

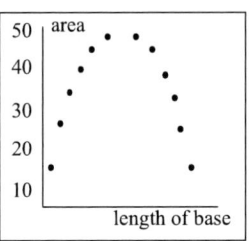

b	h	b × h
1	13	13
2	12	24
3	11	33
4	10	40
5	9	45
6	8	48
7	7	49 (largest)
8	6	48
9	5	45

We find that no matter what value we choose for the perimeter, the greatest product $b \times h$ (which is the area) is found when b and h are equal. However, without more formal algebraic techniques, this generic example is as far as we can get with our *proof*, so at this stage the idea remains a conjecture, although it is supported by strong evidence.

The graph of the area, compared with the base for the sample perimeter of 28, is shown above the chart. Such relations are called *quadratics* (from the term *quadrilateral*) and are generally explored more formally in high-school curricula. The task

> **INSIGHT**
>
> For a given perimeter, the square is the rectangle with the largest area.

at the end of this section provides a more general proof that the square rectangle always has the largest area for a given perimeter. In fact, it does turn out that the conjecture of the square being the largest rectangle for a given perimeter is always the case. For now, let's investigate this concept further with other shapes.

Exploration/Task

Use what you know about the perimeters and areas of squares and circles to determine if a square or a circle with a perimeter of 100 centimetres has a larger area. Make a prediction before you start. Note that determining the area of the circle will require several steps—you will first need to work backwards from the perimeter to find the radius.

Follow-Up and Discussion

You should have found that each side of the square would be 25 centimetres in length, so its area is 625 centimetres squared. Finding the area of the circle is a bit more complicated. You might have worked backwards from the perimeter of 100 and used that measurement as the circumference to then find the radius. When we know the radius, we can find the circle's area by using the area formula and solving for r. (We can choose to approximate π using the value 3.14, which is fairly conventional.) Note that we are using the techniques for solving equations that were discussed in chapter 12.

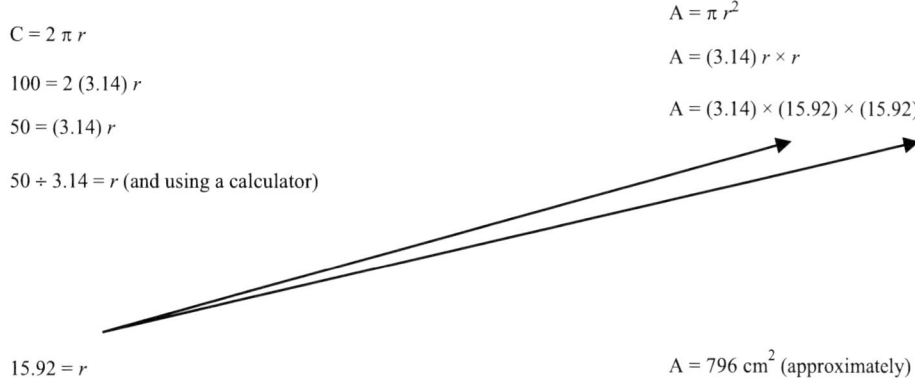

When we compare the two areas, 625 cm² and about 796 cm², we see that the area of the circle is larger than the square with the same perimeter. While this example is not a formal algebraic proof, you could ask each student in your class to repeat the exercise using a different perimeter value and, if their findings align with those above, this would provide a convincing argument.

The conjectured area-perimeter relationship suggests a general idea that applies to many shapes: the idea is that the more compact a shape, the larger the area will be for a given perimeter. A more spread out shape will use more perimeter (outside boundary) for a given area (inside). This is why, for example, it is more expensive to build a house that is very long and narrow than it is to build a house of equal size with a more compact layout that uses less outside wall length.

INSIGHT

The area and perimeter of shapes do not change in direct relation to each other. A more compact shape will have a larger area than a more spread out shape with the same perimeter.

Practice and Further Exploration

1. Explore the conjecture relating area and perimeter by calculating the *area* of the following three regions and comparing them. Assume that each of these regions has a perimeter of 28 metres:
   a) a rectangle with one side that is 2 metres long
   b) a square
   c) a circle
   Which of these three regions was the largest?
2. A rectangle has a perimeter of 12 centimetres. Another rectangle has a perimeter of 14 centimetres. Draw a possible pair of such rectangles, so that the second rectangle (with the perimeter of 14 centimetres) has a *smaller* area than the first one. Label the base and height of each.

Extending the Area Perimeter Reasoning (optional)

Earlier in this current section, a conjecture was made—and strong evidence was provided to support the idea—that a square is always the largest rectangle for a given perimeter. However, such evidence does not constitute a formal proof; further study of this topic is usually done at the high-school level, when quadratics, their vertices, and so on, are examined to fully verify these ideas. It is possible to argue this idea more formally prior to the study of quadratics. The following argument provides an example of a more general approach.

For a given rectangle with a given perimeter P, the base and height added together are always half the perimeter. Thus we know that the sum of base plus height for such a rectangle is a constant. Our conjecture involves the idea that the square rectangle of perimeter P is the largest of all such possible rectangles. A square has one side equal to P ÷ 4, and we will call this x, for simplicity.

x ☐ where P = 4 × x

The area of the square = $(x) \times (x)$

The square rectangle of perimeter P and side x has area x^2. Will this be the largest of the possible rectangles? Recall that base plus height is always a constant—it is half the given perimeter. If we start with a rectangle with a base that is one *more* than x, its height must be one *less* than x, since the sum is a constant:

$x - 1$ ☐ area = $(x + 1) \times (x - 1)$
$x + 1$

If we do this systematically, we can check the next few rectangles in a chart. We need to recall the expansion of an expression like $(x + 1) \times (x - 1)$, which is $x^2 - 1x + 1x - 1 = x^2 - 1$ (see chapter 12).

Base	Height	Area
x	x	x^2 (the square)
$(x + 1)$	$(x - 1)$	$x^2 - 1$
$(x + 2)$	$(x - 2)$	$x^2 - 4$
$(x + 3)$	$(x - 3)$	$x^2 - 9$
and so on		

Following this pattern in general, we see that all of the rectangles that are not square have a smaller area than the square one; therefore, the square must always have the largest area.

14.5 Surface Area

The phrase *surface area* is used to describe areas found on the outside surfaces of closed three-dimensional shapes. For example, if we cut along some of the edges of a cereal box so that we can lay it flat, the sum of all the rectangular areas is the surface area.

> **MATHEMATICAL TERM**
>
> The *surface area* of a three-dimensional object refers to the total (two-dimensional) area of its outside surfaces.

We can determine the surface area of any shape composed of areas for which we have the measurements; ideally, we will also have a method or formula for finding the area. Visualizing the outside surface of an object as a collection of flat shapes is often the main challenge when determining surface area. When a three-dimensional object is cut apart and laid flat, the resultant shape is called the *net* of the shape. There is generally more than one net for any given object. For example, for the box pictured here,

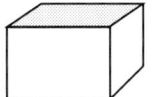

there could be various nets, for example,

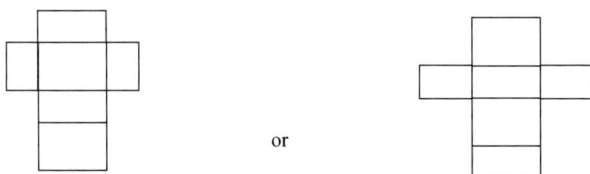

> **MATHEMATICAL TERM**
>
> The *net* of a three-dimensional shape is a two-dimensional shape that can be folded along its edges to create the three-dimensional object.

Getting students to cut up cardboard containers is a great way to help them visualize surface area and find the nets. One shape that can be a little more challenging to work with is the (right) cylinder. When we cut apart a cylinder, we get two circles and a rectangle. (An easy way to demonstrate this for students is simply to remove the label, base, and lid from a soup can. You can also use a cardboard toilet paper or paper towel tube, using paper and tape to add circular ends.)

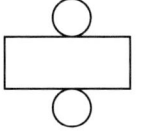

To find the area of the two circles, we need to know their radius, and then we can use the circle area formula for each. This radius is the same as the radius of the cylinder. But what about the area of the rectangular label? The rectangular part of the net (such as the soup can label) is called the lateral surface area. We can see that the height of the rectangular label, or lateral surface area, is simply the height of the cylinder. To find the area of a rectangle, we also need to know the length of its base—this is trickier.

Exploration/Task

Explore your cardboard tube or label by switching it back and forth from a cylinder to a rectangle. Look for a way to calculate the base of the rectangular label, given your knowledge of the cylinder's radius.

Follow-Up and Discussion

When we cut open a cardboard tube and lay it flat, we see that the length of the base of the rectangle formed from the surface is exactly the *circumference* of the original circular tube—and we have a formula for calculating circumference (see section 14.1). The total cylinder surface area—the two circles plus the rectangle, can be expressed as:

SA = 2 circle areas + 1 rectangle
SA = 2 × π × r × r + (base, which is the circumference of the cylinder) × h
SA = 2 × π × r × r + (2 × π × r) × h

This formula is sometimes simplified to SA = $2\pi r^2 + 2\pi rh$. There are other versions of this formula in which the terms are rearranged, but because they tend to obscure the connection to the construction of the formula from the net, we do not recommend using them.

Practice and Further Exploration

A tin can has a diameter of 6 centimetres and a height of 12 centimetres. Draw a sketch of the net of the amount of tin required to manufacture the can. Label all of the dimensions, and calculate individually the area of the two circular ends, as well as the rectangular part (the lateral surface area). Find the total surface area.

14.6 Volume

Volumes of Right Shapes

> **CONNECTION**
>
> To review the names and features of three-dimensional shapes, see section 13.4.

The volume of *right* (meaning that if they stand vertically, they can be seen as composed of equal horizontal layers) prisms, cylinders, and so on can be determined using a common conceptual approach.

Example

A great way to begin exploring volume is to start with a few small boxes, such as small chocolate or jewellery boxes that are rectangular prisms. Fill a box with centicubes, as pictured below:

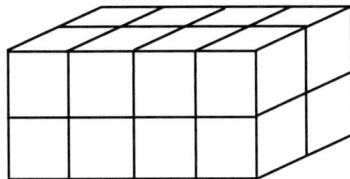

The bottom layer of centicubes in the box has a surface whose area is the same as the area of the *base* of the box. The method for calculating the area of the base should be familiar to students, as it is just a rectangle with an area of base times height. In the above diagram, this area is 2 × 4 or 8. The total number of cubes in the box is the volume—the total space in the box. We can determine the volume by calculating the number of cubes on the base multiplied by the number of layers, which is the height of the box. In other words, we could represent the volume of the box by placing 2 layers of 8 cubes in the box. Here is one such layer:

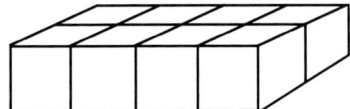

Reasoning

The area of the base tells us the number of centicubes in the bottom layer. The height tells us the number of such layers. It doesn't take many more examples for students to see that the volume of a rectangular prism (a box) is always the area of the base multiplied by the height. We conclude that the volume of rectangular prism = *area of base* × *height*. While this formula is sometimes written as $l \times w \times h$, we find the above version, area of base times height or simply $b \times h$, often generalizes more easily to other shapes. In fact, the formula generalizes to any right shape (shapes whose horizontal layers are identical). For example, for this triangular prism,

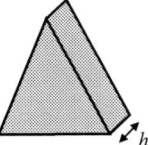

as long as we think of the *back* of the shape, as pictured here, as the base,

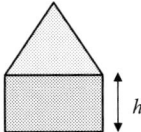

we can see that it is composed of identical triangular layers. As long as we know the area of the triangular face (now the base), as well as the height h as marked, we can find the volume of the triangular prism by multiplying the area of the triangular base by the height (which is the number of layers). The base-times-height volume relationship works with any right 3-D shape, as long as we imagine standing it on the face that is the base, and counting identical flat layers. For a right cylinder, such as the one pictured here,

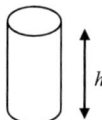

as long as we know the area of the base—for example, using the circle area formula—we only need to know the number of circular layers, or the height, to calculate the volume.

> **CONNECTION**
> To review circle area, see section 14.3.

Reasoning

We see that we can find the volume of any right shape by multiplying the area of the base by the height—because the height is simply the number of these layers. We can write the formal version of the cylinder volume formula as follows:

$$\text{Volume} = \text{area of base} \times \text{height}$$
$$= \pi r^2 \times h \text{ (because the base is a circle)}$$

It is almost unnecessary for students to remember this formula once they understand the concept of volume as the area of the base multiplied by the height. They can easily use reasoning to apply this idea to any right shape.

Units of Volume

Students who learn about area and volume without sufficient attention to tangible objects do not typically have a visual understanding of the units. This connects with the conceptual difficulties described earlier in this chapter with regards to conversions (see sections 14.1 and 14.2 in particular). Physically covering a rectangular area with 1-centimetre-by-1-centimetre *square* tiles provides a visual image of *square* centimetres (cm²), while *filling* a rectangular object with 1-centimetre-by-1-centimetre-by-1-centimetre *cubes*—which are appropriately called *cubic* centimetres (cm³)—provides a visual image of a unit of volume. Because there are now three dimensions to consider, determining the units can be even more challenging.

Example

Place a centicube and a thousands cube (block) from a set of base ten blocks in front of you and compare them.

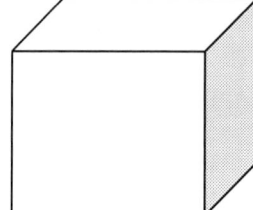

The block is 10 centimetres in each direction—and 10 centimetres is a decimetre; therefore, the block is called 1 *cubic decimetre*. This is a unit of volume that is equivalent to a cube with edges of 1 decimetre. Its dimensions are 10 centimetres by 10 centimetres by 10 centimetres, so it represents or holds $10 \times 10 \times 10$ or 1,000 *cubic centimetres*, each of which has edges of 1 centimetre. As we saw previously in this chapter, using models when converting units is very important.

Practice and Further Exploration

1. Measure the outside dimensions of a food box such as a cereal box. Use the measurements to calculate the volume of the box. Imagine the number of centicubes that would be required to fill the box.
2. Measure the dimensions of a pop can and use them to calculate the volume of the can. Note that 1 cm³ of volume (think of this as a 1-centimetre-by-1-centimetre-by-1-centimetre cube of volume) is also called 1 millilitre (ml). Compare your answer with the stated volume of the contents of the pop can.

14.7 Mass and Capacity

The metric system is designed to exactly connect mass (weight) and volume, based on working with water. If we imagine filling a thousands cube or block from a set of base ten blocks with water, we would need 1,000 millilitres, or 1 litre, of water. The water would weigh 1 kilogram. Working backwards, 1 cubic centimetre of water weighs one one-thousandth of a kilogram, which is 1 gram (recall that *kilo* means one thousand), and since it is one one-thousandth of a litre, it is also called a millilitre. A millilitre of water weighs 1 gram.

Working with the basic relationships described in the previous paragraph and a set of base ten blocks to represent the various amounts is the best way to visualize these amounts and the relationships among them. No set of formulas is required to understand conversions from one unit to the next; rather, the basic definitions, together with models and reasoning, allow us to navigate the ideas.

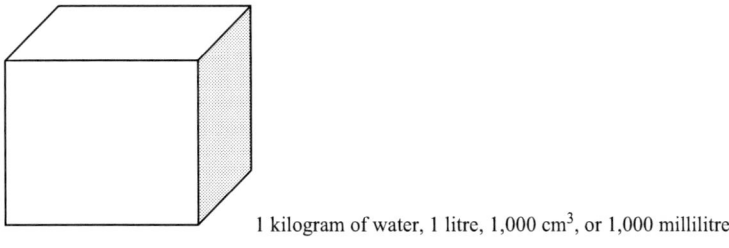

1 kilogram of water, 1 litre, 1,000 cm³, or 1,000 millilitres

1 gram of water, 0.001 litre, 1 cm³, or 1 millilitre

Practice and Further Exploration

Convert 2,310 millilitres to litres. If this were water, what would it weigh? Express the weight in both grams and kilograms.

Chapter Problems

1. a) If you put your arms around a large cylindrical garbage can, you can just touch your fingertips together. Measure your arm span, and use this measurement to determine the diameter of the garbage can.
   b) If your friend did the same thing with a different garbage can, and his arm span was 6 centimetres less than yours, what would be the approximate diameter of the second can?
2. One square is 10 centimetres by 10 centimetres, and another is 30 centimetres by 30 centimetres.
   a) Predict how many times bigger the area of the second square is than the first.
   b) Calculate each area and check your prediction.
   c) Draw a model of a square metre, and use your model to convert each of the areas of the two squares to square metres.
3. Find a soup can or other food can. Measure the can and calculate the volume in cubic centimetres. Examine the volume listed on the label, and compare your answer to this amount.

Further Reading

Callahan, K. M. (2011/2012). Listening responsively. *Teaching Children Mathematics, 18*(5), 296–305.

>Callahan discusses helping pre-service teachers understand concepts of mathematical discourse through the lens of discussing relationships in area and perimeter. The article focuses on actual discussions about the mathematics associated with area and perimeter, and the relationships between the two concepts.

Colgan, L. (2012). Not-quite-so elementary mathematics: The mathemagic of trees. *Ontario Mathematics Gazette, 51*(2), 12–14.

>Colgan examines different types of triangles by measuring the height of trees, and uses circle circumference to determine the approximate age of the trees.

Dietiker, L. C., Gonulates, F., and Smith III, J. P. (2011). Understanding linear measure. *Teaching Children Mathematics, 18*(4), 252–259.

>This article discusses tasks used to teach students about linear measurement and possible enhancements to make the tasks more conceptual. The authors discuss the need for more hands-on conceptual problems in linear measurement to allow students to understand underlying concepts, so that errors and misconceptions are less likely.

Kajander, A. (2012). MB4T … When is "close" good enough?: The circle area formula. *Ontario Mathematics Gazette, 51*(2), 22–23.

>The author concretely links the building of the circle area formula to the deconstruction of a circle.

Siegel, A. A., and Ortiz, E. (2012). Perimeter and beyond. *Teaching Children Mathematics, 19*(1), 38–41.

>Siegel and Ortiz use a perimeter question that forms the basis for a discussion the relationships between various mathematical concepts, including addition, subtraction, geometry, and reasoning.

Chapter 15
Data Management and Probability

15.1 Introduction to Data Management

The need for the relatively new mathematical strand of data management arises from our technological and information-based society. Citizens are expected to understand, for example, what is implied if we say that 0.2% of the population might get a certain disease, or that the mean house price in a particular area is $400,000. It is difficult for the human mind to conceive of very large numbers. For example, imagine a large park as a venue for a rock concert. Would you be able to tell if there were 2,000 people there? Or 5,000? Or 10,000? What does it mean to say that a country is a million dollars in debt? Or a billion? How much bigger is one than the other? This difficulty in truly understanding large numbers can also make it challenging to understand world-related data.

Techniques of data management are meant to give students some tools to represent and compare data, which will allow them to better and more usefully understand such data. These tools include graphical techniques and, eventually, statistical methods. For the purposes of elementary education, the focus is on graphical representations, with only very rudimentary statistical techniques involved. Since the purpose of data management is to help us better understand our world, there is no other curricular mathematical strand that is more amenable to interdisciplinary study. In fact, we suggest that the majority of data management lessons be connected with the context of other subjects, such as social studies. There is also some movement in mathematics education to include social justice as a theme, when feasible, in mathematics lessons. Data management is a particularly useful strand for introducing such themes. We will begin our discussion of data management by looking at what types of data might be encountered, and how such data can be conceived mathematically.

15.2 Discrete and Continuous Data

> **MATHEMATICAL TERM**
>
> *Discrete data* is numeric data that cannot be subdivided beyond a certain point. There are gaps between the numbers.

> **MATHEMATICAL TERM**
>
> *Continuous data* involves numbers with no spaces or gaps between them. When working with continuous data, it is always possible to find another value between two numbers.

The use and characterization of groups or lists of numbers that have been isolated from any context can be misleading. You may recall the everyday term *the average*, which is used as a way to describe a typical value of a set of numbers. For example, we might calculate the average of the numbers 2, 5, 4, and 3 and get the value 3.5 (recall the numeric method of adding the numbers and dividing by how many there are). Mathematically, 3.5 is called the mean value. However, without a context, this information has no meaning. If, for example, we are trying to predict the number of people who will attend an event based on the attendance of 2, 5, 4, and 3 people at four previous events, then a mean of 3.5 doesn't make literal sense. When counting people or any other quantity that can't be divided up indefinitely, we say the data is *discrete*; discrete data cannot be subdivided beyond a certain point.

It is possible to have discrete values that are not whole numbers. For example, if pie is sold in pieces of size $\frac{1}{6}$, the total amount of pie sold might involve a fraction, but the total will change by amounts no less than $\frac{1}{6}$ (or in multiples of $\frac{1}{6}$).

Continuous data, on the other hand, represents amounts that can be subdivided as many times as we like, or *infinitely*, as a mathematician might put it.

The concepts of discrete and continuous sets of numbers were also discussed in chapter 10. Concrete examples of continuous quantities include distance, or a length of fabric. At times however, the units used to describe a continuous quantity are discrete, which may cause some confusion. This ideas will be explored in the following task.

Exploration/Task

Consider the following discussion between two students. Jake and Stu are discussing experimental data that represents time. Jake believes that the data is continuous because time doesn't have gaps. Stu believes that the data is discrete because it is recorded in intervals of minutes and seconds, based on the timer used for the experiment. If possible, discuss these ideas with a colleague, and think about the issues raised here.

Follow-Up and Discussion

Jake is correct in saying that the data itself is continuous; however, the measurements are in discrete intervals, as Stu recognizes. Such data could be represented on a bar graph showing intervals, but with no gaps or space between the bars, which suggests continuity. This type of bar graph is actually called a histogram. There is more to come on this in the following section.

15.3 Sampling and Representing Data

As mentioned, no strand of elementary mathematics is more amenable to interdisciplinary study than the collection, representation, and interpretation of data. Students often have very little connection to data that does not interest them, or touch their world in some way. Although data about student attributes—for example, the number of pets they have, or their heights—can be collected, examining such data may not be particularly exciting or illustrative to students. Alternatively, data about issues related to social justice and world events, such as hunger, child labour, and so on, may bring particular meaning to these lessons, while also allowing students to study other subjects such as social studies and geography.

Sampling

A *sample* of data is a collection or cross-section of data from a larger data set from which it might be difficult to collect every single piece of data. An opinion poll, for example, gathers the opinions of a certain number of people and assumes that this data represents all members of a specified larger group. A representative sample is one in which the sample reasonably predicts characteristics of a larger group. In contrast, a biased sample does not. For example, one might sample the first 100 people to arrive at a mall food court in the morning by surveying them about their favourite brand of coffee, and then use that data to generalize to a larger population. But if the food court has only one coffee shop, the people arriving early to the food court might be doing so specifically because they prefer that brand of coffee; if they liked another brand, they may instead go to the mall down the road. Trying to anticipate situations that can lead to a biased sample is a central challenge to collecting a good sample.

Bar Graphs and Histograms

Two of the most common ways to represents sets of data are *bar graphs* and *histograms*. These two methods are very similar, except that bar graphs are used to show discrete data, by separating the bars, while histograms are used to represent continuous data, which is illustrated in *intervals*.

Example 1: Bar Graph

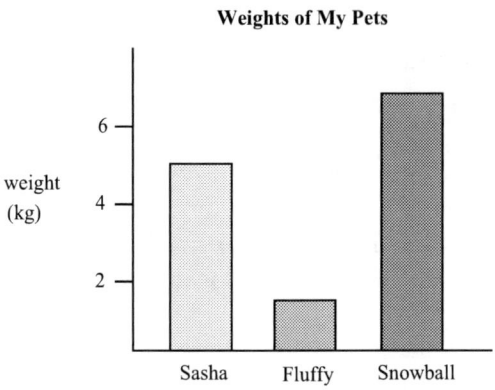

Example 2: Histogram

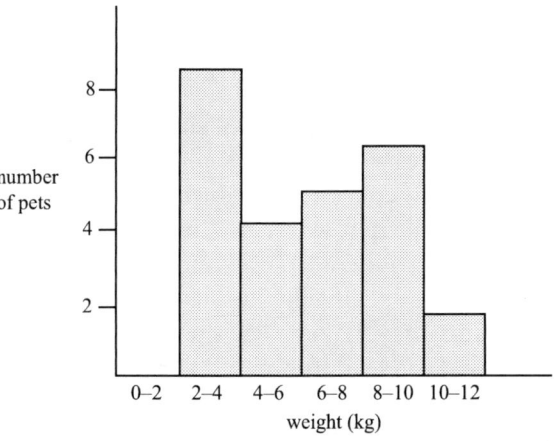

In the histogram, the edges of the bars touch. A value right on the edge of two intervals is generally counted in the higher category; for example, a dog weighing 10 kilograms would be counted in the 10 to 12 kilogram bar.

Other kinds of bar graphs that use small pictures to represent a number of items are called *pictographs*. They are popular in the media; look for some next time you read a magazine or newspaper. Simple pictographs also appeal to young learners.

Scatter Plots and Line Graphs

A dot can be used to represent each individual data item, in a manner similar to the way points are graphed (see chapter 12). These points can be joined by individual line segments to further show or illustrate the connection between them. An estimated line or curve is sometimes drawn that approximates the points, but doesn't necessarily go through all of them, and suggests a general relationship. Such estimated lines and curves are called lines of best fit and curves of best fit, and are often also discussed algebraically in the context of early high-school level algebra.

Stem and Leaf Plots

While charts of numbers might allow all of the included values to be visible, getting a sense of the overall data from such lists can be difficult if they include many numbers. On the other hand, tallies, bar graphs, and other types of frequency tables or graphs tend to show data in intervals, and the actual values are no longer visible. Such methods can help users get an overall sense of large lists of data, but not of the individual values within the data.

One technique that has some elements of both methods is to separate the leading digit or digits—the biggest part of the number—and use it to determine the organizational categories (called the stem), while showing the smaller digits as individual items in that category (called the leaves).

Example

Imagine that the following data was collected about weights of pets owned by members of a class:

1.3	1.5	1.6	1.8	1.8	1.9	1.9
2.1						
3.2	3.5					
4.4	4.5	4.7				
5.8						
6.2	6.7	6.9				
7.5	7.8					
10.0						
11.1	11.6					

If this data were represented in a stem and leaf plot it would appear as follows:

Stem and Leaf Plot

Weights of pets in kilogram

1.	3	5	6	8	8	9	9
2.	1						
3.	2	5					
4.	4	5	7				
5.	8						
6.	2	7	9				
7.	5	8					
8.							
9.							
10.	0						
11.	1	6					

We see that the stems for this data are the whole number parts. The smaller parts, in this case the values to the right of the decimal, are the leaves. The leaves extend to the right, showing the decimal part of each number with the same stem on the same line. Practice reading the stem and leaf plot by comparing the plot to the list of numbers. Note that the leaves do not need to be decimal parts of numbers—for example, a list of numbers such as 101, 105, 108, 108, and 109 can be represented with a stem of 10 and leaves of 1, 5, 8, 8, and 9.

Pie Charts

Pie charts are commonly found in journals and pamphlets as they are relatively easy to read. For example, a charity might use a pie chart to convince you that most of your donated money goes to the program you are supporting, rather than to administration. One caution, however, is that while pie charts are relatively easy to read, they are not as easy to construct, especially with real-world data. This is important for teachers to remember when asking students to construct pie charts by hand using data that they have not prepared. There are a number of great websites that will create a pie chart for students when they simply enter the numbers. This is a great idea for students whose computational skills may make constructing pie chart wedges based on, for example, 68%, 4%, and 38% very difficult.

Learning to determine the angles of the pie pieces without software is quite challenging, and makes a wonderful extension activity for a more able child; it doesn't become part of the curriculum until later grades. This work involves knowledge of multiple concepts, including fractions, ratios, and measurement of angles. At many grade levels, it is sufficient to be able to read and interpret pie charts, and to construct only those charts that include simple data that result in straightforward fractions without software.

Example

The following pie chart shows that in a class of 20 students, 10 people got a grade of B, 5 people got a C, and 5 got an A:

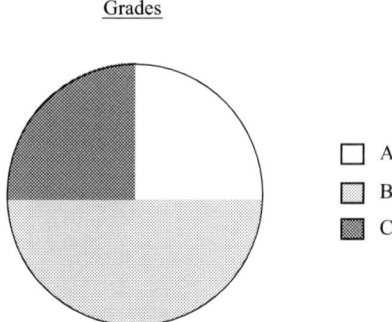

This pie chart would be reasonably easy for children to construct because the numbers in the data form fourths of the total 20. On the other hand, representing 70 and 30, for example, on a pie chart is more difficult. In this case, a chart divided into approximately $\frac{2}{3}$ and $\frac{1}{3}$ is a good start, but is not quite accurate. For the total represented here, 30 + 70 or 100, the $\frac{70}{100}$ amount is not exactly $\frac{2}{3}$. How might the angles of the wedges be determined without using software if accuracy is required?

Each value in the pie chart represents a fraction of the total; therefore, the total is the sum of all the values. In the example of 70 and 30, the total is 100. The region representing the 70 should be $\frac{70}{100}$ or $\frac{7}{10}$, and the region representing the 30 should be $\frac{30}{100}$ or $\frac{3}{10}$. In this case, the numbers were fairly simple, and we could simply use tenths pieces from a set of fractions circles, but this is not always the case.

When determining the angles of the wedges in a pie chart, the idea that the total circle rotation is 360 degrees is important. Each fraction of the total—in this case, $\frac{7}{10}$ and $\frac{3}{10}$—can be used to find the fraction of 360 degrees for the angle of that portion of the pie chart. The larger region should have an angle of $(\frac{7}{10}) \times (360)$ or 252 degrees, and the smaller region should have an angle of $(\frac{3}{10}) \times (360)$ or 108 degrees. To double-check the answer, we can add the two angles. We find that the sum of 252 + 108 is the full 360 degrees. In such cases, a protractor—or dynamic software—is required to draw the angles in the pie chart. You may recall that a protractor is a tool that allows the measurement of angles. Students usually need significant support when learning to use manual protractors.

15.4 Measures of Central Tendency

A central goal of data management is to give us ways to represent and describe complex data so that we can better interpret it. Statistics is the branch of applied mathematics that relates to such analysis of data. The following are examples of statistical questions: How large a sample is needed to make a conclusion about a population? What is a typical value? How spread out are the values? Is there a trend? Are there relationships within the data?

Ways to describe a data set or list of numbers can involve several types of measures, including measures of dispersion and measures of central tendency. Measures of dispersion tell us how spread out a list of numbers is. For example, a group of students whose grades are 45, 52, 67, 75, and 90 will likely be more difficult to teach than a group of students whose grades are all in the 60s. A study of such measures of dispersion is not generally included in the elementary curriculum. Measures of central tendency, on the other hand, are usually addressed at the elementary level; these measures include mean, median, and mode. These numbers tell us something about the typical value of a data set; however, students are often simply told how to calculate each of these measures, and then asked to practice computing them. Students must learn that each of these measures has a different use, and that, at times, not all of the measures are appropriate for certain data sets. Understanding when and how to use each measure is important.

The *mean* (for which the word *average* is sometimes informally used) is the most common of these measures. Many of us learned the mean through the computational definition of *add up all the numbers and divide by how many there are*. This definition does not illustrate the central notion of the mean, as we will explore below.

Exploration/Task

1. A box of 20 pencils is distributed to 5 students, with the students getting 1, 2, 4, 6, and 7 pencils respectively. The students decide that this is unfair,

and pass around pencils until everyone has the same number. How many pencils does each student end up with?

2. Calculate the mean of the list of numbers—1, 2, 4, 6, and 7—provided in the previous question. How does this value compare to the number of pencils each student has after the redistribution? Noticing this, and knowing there were initially 20 pencils, is there another way that the mean could have been determined? What might the mean really tell us about the 20 pencils and the 5 people?

Follow-Up and Discussion

From the task, we can see that the mean is actually the *fair share*. As another example with a social justice theme, the mean wealth per person on the planet is the amount everyone would have if the world's wealth were distributed fairly, that is, all the world's wealth collected and then redistributed so that everyone got the same amount. The mean would be greatly affected by someone who is very rich—or by large numbers of people who are very poor.

The median, on the other hand, is less influenced by a few values that are greatly out of alignment with the others. The method used to calculate the median is to determine the middle number in a list of values that is written in order of size. If there is an even number of values, the midpoint of the two middle values is taken. Let's compare the median and the mean for a set of data.

Exploration/Task

Weekly net salaries in a company of six people are $1,000, $1,200, $1,500, $1,600, $1,700, and $8,000. Compute the mean and median salary. Which tells more about the typical salary? Next, assume that the president, who makes $8,000, takes a pay cut and will now receive $7,000. Recalculate the mean and median. What do you notice?

Follow-Up and Discussion

You should have noticed that while the mean is influenced by the president's pay cut, the median is not. The mean in the initial scenario is $2,500. We observe that this value is larger than everyone's salary except the president's, so it is a bit misleading in terms of portraying a typical salary. The median in both scenarios is the midpoint between $1,500 and $1,600, or $1,550. This amount may seem a more accurate typical value for this data than the mean, and it remains the same when the president's salary is reduced. The mean, in contrast, is pulled towards an extreme (meaning far from typical) value, but also changes when an extreme value is changed. It is important to realize that while the mean is often used to typify data, it can be unreasonably influenced by extreme values, thus providing an unrealistic picture.

Care should also be taken when interpreting mean and median of data that represents only whole numbers. While some data, such as that involving money, temperature, distance, and so on, makes sense as partial units (for example, *half* a kilometre), other data, such as the typical number of people with a certain characteristic, does not. The mean (and median) number of people of the list 1, 2, 3, and 4 is 2.5. While this doesn't really make literal sense, statisticians would think of this result as implying that 2 or 3 people is equally likely.

The third commonly used measure of central tendency is the mode. The mode tells us the most frequent or likely response. It should be noted that situations for which the mode is useful tend to differ from those for which mean and median apply; therefore, simply asking students to calculate the mean, median, and mode for lists of numbers misses the point of what is important to learn about these measures, which is to determine which one to use in a certain situation. We will examine the importance of context in the following task.

Exploration/Task

The cafeteria has three lunch specials as follows:
Choice 1: hamburger and fries
Choice 2: macaroni and cheese with salad
Choice 3: cheese, crackers, and fruit plate
The first 10 people in line choose the following specials:
   2   3   1   2   1   3   2   2   3   3
Find the mean, median, and mode for this data *without reference to the context*. Now return to the context. Which measures make sense to describe this data? Which do not?

Follow-Up and Discussion

The context illustrates a different kind of data than we looked at previously. The numbers 1, 2, and 3 do not tell us anything about the relative *size* of the lunch specials. Rather these numbers are called *nominal* numbers—they simply separate the data into categories. Computing the mean and median is not helpful—a mean of 2.2 doesn't tell us anything useful about this context as there is no special 2.2; the same problem exists with the median. It doesn't make sense to line the values up in order of magnitude as no magnitude is implied. On the other hand, this type of data is well characterized by the mode, as we are indeed interested in the most likely choice. We see that this particular data set contains equal occurrences of the specials 2 and 3, hence, we say the data is *bi-modal* because there are two modes, 2 and 3. This tells us that specials 2 and 3 are the most popular and are equally likely to be chosen, at least in this group of 10 people.

In general, care should be taken to align contexts and measures appropriately, and to interpret what the measures tell us. Using such measures can make data analysis easier, but our own reasoning and interpretation are important, too. When students are asked to design their own surveys, it is important to encourage them to think about which measures they will use to analyze the data, and why.

15.5 Concrete Models of Probability: The Draw Ticket

Many of us have wondered if it is better to pay $10 for a draw ticket for a million dollars, or $2 for a 50/50 draw of about $500 at hockey game. A key point in determining the likelihood of winning a prize is *how many tickets* are entered in the draw.

Exploration/Task

For the above example, assume there were 200,000 tickets sold for the million-dollar draw, and 500 tickets sold for the hockey draw. Does this help you decide which is a better bet?

Follow-Up and Discussion

These kinds of real-world examples may be great conversation starters for discussions about probability and chance—but there is no "best bet" answer in this task. What is a 1-in-200,000 chance? Is a 1-in-500 chance of winning $500 a better bet? When numbers are this large, it is difficult to really make sense of the situations.

Simple classroom experiments using more manageable numbers provide an easier starting point. For example, in a classroom game, rolling a number cube (a 6-sided die, for example) could result in any of the numbers 1, 2, 3, 4, 5, or 6. This list of numbers (the possible outcomes) is called the *sample space*.

Example

Think about the following with respect to rolling a six-sided number cube (die):

- A roll of 6 wins the game. What are your chances of winning?
- What is the chance that you will roll an even number?

If the die has 6 faces, you can expect that $\frac{1}{6}$ of the time you will roll a 6. In relation to the second question, 3 out of the 6 faces contain even numbers, so you can expect to roll an even number about half the time. These two scenarios help us understand the chance of a particular event happening, which is called the *probability*.

$$\text{The probability of an event is } \frac{\text{\# of occurrences of the event}}{\text{\# of events in the sample space}}$$

For example, the probability of rolling a 2 on a single roll of the number cube is $\frac{1}{6}$—1 occurrence out of 6 possible occurrences in the sample space. One of the things to look for when deciding whether to buy a lottery ticket is the number of tickets sold. In our example, 200,000 tickets were sold—that's a lot of tickets!

15.6 More than One: Dice Rolls

Predicting probability is more complicated if there is more than one event occurring. A pair of number cubes (such as 6-sided dice) can be used to provide a manageable classroom example of this. If your students are playing a game that requires them to add the sum of two dice rolls, they may initially assume they are equally likely to get any number from 2 to 12. (Initially, they may even think they could roll sums with any of value from 1 to 12.)

Creating a list of the sums obtained from repeated dice rolls—perhaps by having everyone in a group record their own rolls, and pooling the data on one tally chart—demonstrates that some numbers are more likely to occur than others.

If you imagine the dice are two different colours, and create the sample space for each die, some sums appear more frequently than others. The sample space for the sums in this experiment is:

	Red die	1	2	3	4	5	6
Green die	1	2	3	4	5	6	7
	2	3	4	5	6	7	8
	3	4	5	6	7	8	9
	4	5	6	7	8	9	10
	5	6	7	8	9	10	11
	6	7	8	9	10	11	12

Since the probability is the number of occurrences of a particular outcome divided by the total number of the possibilities, we see that the probability of rolling a sum of 12 is $\frac{1}{36}$; however, the probability of rolling a 7 is $\frac{6}{36}$ or $\frac{1}{6}$—a much greater chance than getting a 12. No wonder some people think 7 is a lucky number!

Exploration/Task

Consider the following example, and draw a diagram to show the possibilities. The school cafeteria offers several different kinds of sandwiches. They are on either white or brown bread. There are three different sandwich fillings: tuna salad, ham and cheese, and cucumber and tomato. Each sandwich is available on both types of bread. How many choices are there in all? If someone was going to buy you a sandwich without knowing your taste, what is the probability they will get your first choice of both bread and filling if they chose randomly?

Follow-Up and Discussion

One way to show the combinations is a tree diagram.

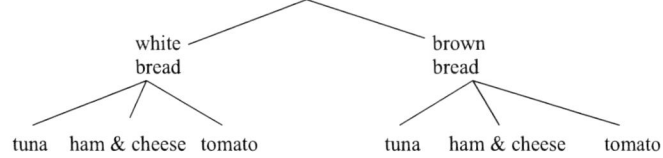

Here there are 6 choices in all. The chances of, for example, randomly picking tuna on brown bread is 1 in 6. We say the *probability* is $\frac{1}{6}$.

Exploration/Task

Imagine you went on a trip, and packed 4 shirts and 2 pairs of pants. Each pair of pants can be worn with any of the shirts. How many outfits do you have? Try to first make a prediction without drawing a tree diagram, then draw the diagram to confirm your answer.

Follow-Up and Discussion

For each pair of pants, there is a choice of 4 shirts—so there are 4 × 2 or 8 total outfits to choose from. You may have noticed that the number of combinations (which is really the *sample space* of a probability experiment, meaning all of the possible outcomes) is the *product* of the number of choices. For example, in the sandwich problem, there were 2 × 3 choices of sandwiches. Returning to the first example in this section, with the two colours of (6-sided) dice, we see that the sample space has 6 × 6 possible outcomes; we must remember that each total roll is really the sum of the scores on each of the two different coloured dice. The number of choices is important because finding the probability of *one* of these outcomes requires knowing how many possibilities there are in total.

Practice and Further Exploration

1. A board game uses both a 6-sided die and a coin. According to the rules, you get an extra turn if you roll a 6 and flip a head on the coin. What are the chances of this happening in any given turn?
2. In the game described above, the rolls you made during your first three turns did not give you an extra turn. Are you more likely to roll a 6 and flip a head on your fourth turn, given your first three turns did not yield this combination?

15.7 How Many Ways? Full Circle Back to the Number System

Many lottery numbers have a significant number of digits. It is really difficult for humans to concretely appreciate the difference between very large numbers such as one million and one billion. In fact, one billion is 1,000 times greater than one million! Similarly, you would be 1,000 times as likely to win a lottery with 1,000 choices as one with 1,000,000 choices, if you bought the same number of tickets for each.

Exploration/Task

Imagine you work for a phone company and you need to plan phone numbers. A new local (seven-digit) phone number zone will have the first three digits of 863. Assuming for now that *no* numbers will be excluded, how many new phone numbers will there be?

Follow-Up and Discussion

This problem can be thought of equivalent to asking how many four-digit numbers there are. Using our knowledge of combinations, we see that there are 10 choices for each digit (the numbers 0 to 9), so there are 10 × 10 × 10 × 10, or 10,000, choices in all. You might at first wonder why this isn't 9,999 rather than 10,000, if you noted that the digit combinations range from 0000 to 9999. The number of values between 0001 and 9999 is 9,999, but this does not include 0000. There are 9999 + 1 or 10,000 choices in all, as the following table shows:

Choice number	Phone number
1	863-0000
2	863-0001
3	863-0002
4	863-0003
.	
.	
.	
9,998	863-9997
9,999	863-9998
10,000	863-9999

We can observe that the digits in our number system—10 choices in all for each digit—directly predict how many values can be represented! This interesting observation opens the door for study of bases other than base ten. There are 10 symbols available for each digit in our decimal system. Quantities represented in a different base than ten behave differently because there are a different number of choices for each digit. As an example, the following table illustrates how numbers are composed in base five. Notice the relationship of 25 and 5 times 5.

Decimal number	Representation in base five
1	1
2	2
3	3
4	4
5	10
6	11
7	12
8	13
9	14
10	20
11	21
.	
.	
24	44
25	100

Similarly, the next number after 7 in base eight is 10_8 (the subscript indicates this is a value in base eight) and after 77 is 100_8. While the topic of other bases is outside the range of many elementary curricula, it makes for an interesting extended research topic for students, particularly gifted students who might be in need of challenge.

Practice and Further Exploration

1. A radio station runs a contest for listeners—if you can guess a seven-digit phone number correctly, you will win a large sum of money. The station tells callers, "you only have to guess seven numbers!" Assuming that the first digit can't be zero, what are your chances of guessing the correct seven-digit phone number?
2. Insurance companies use particular kinds of mathematicians called actuaries to determine the likelihood of certain events happening, and thus help them set prices. Consider the simplified scenario in which, in one year, for every 1,000 customers an insurance company has there will be one insurance claim paid out in the amount of $100,000. What is the net cost per year in claims of the insurance service *per customer*? If the company wants to make a 50% profit, what should it charge each customer per year for this particular insurance plan?
3. Write out the first 10 numbers in base three. What decimal number is represented by the base three number, 100_3? What is 100_3 in base ten? What relationship does this latter value have to a product of threes in base ten?

Chapter Problems

1. Some teachers may feel that real-world data, particularly data related to social justice, has no place in classrooms; however, the websites of some provincial education ministries now have links to information about social justice–based mathematics lessons for teachers.
   a) Read the short article, "Putting 'Pizza Party' Math to Rest," written by an Ontario teacher, available at: http://www.policyalternatives.ca/publications/monitor/february-2005-putting-pizza-party-math-rest
   b) Construct a classroom problem using the concepts of this chapter and the theme of social justice. Solve your own problem.
2. a) According to the Winter 2009 issue of *Animal Talk*, a publication of the Toronto Humane Society, the National Shelter Statistics for the Canadian Federation of Humane Societies show that 6,962 out of 36,651 dogs were euthanized by their affiliates in 2008. (Note that this does *not* include data from municipal pounds, where rates can be as high as 50%.) Express the *probability* of euthanasia for a dog at the Canadian Federation of Humane Societies as an exact fraction, and then as an estimated fraction using a whole number denominator of 10 or less.
   b) A recent report on the Animal Rescue Site (www.theanimalrescuesite.com), states that of the 26 million unwanted dogs in the United States each year, 10 million are euthanized. How does this compare to the situation

in Canadian humane societies according to the data provided in a)? (Remember that this data does not include municipal pounds.) Create one pie chart for the Animal Rescue Site data and a second pie chart for the Canadian Humane Society data. Show all your calculations with explanation.
3. Look through a recent newspaper or magazine. Find an article that uses graphical data to argue a point. How would you change the graphical representations used if you were going to argue the opposite point of view? Choose a graph or representation in the article and re-represent the data to show this opposing view. For example, you might change the scale of a graph to make the rate of change appear greater or smaller.

Further Reading

Bush, S. B., Karp, K. S., Popelka, L., and Bennett, V. M. (2012). What's on your plate? Thinking proportionally. *Mathematics Teaching in the Middle School, 18*(2), 100–109.

> The authors use the context of healthy eating as a basis for a data management unit to explore different graphs and build understanding of concepts related to proportional reasoning and data analysis. The article also includes reproducible activity sheets to be used in a classroom related to the topic.

Hudson, R. A. (2012/2013). Finding balance at the elusive mean. *Mathematics Teaching in the Middle School, 18*(5), 300–306.

> Hudson shows technology and experience can be used in combination to help students develop a concept of mean.

O'Dell, R. S. (2012). The mean as a balance point. *Mathematics Teaching in the Middle School, 18*(3), 148–155.

> O'Dell focuses on the use of a ruler as a balance to help students understand the concept of mean. She presents different tasks for students to complete to increase their knowledge of the mean.

Whitin, D. J., and Whitin, P. (2012). Making sense of fractions and percentages. *Teaching Children Mathematics, 18*(8), 490–496.

> Whitin and Whitin discuss using the context of advertising during television programs to help students build an understanding of percentages and fractions through the use of pie charts.

References

Ambrose, R. (2004). Initiating change in prospective elementary school teachers' orientations to mathematics teaching by building on beliefs. *Journal of Mathematics Teacher Education, 7*(2), 91–119.

Askey, R. (1999). Knowing and teaching elementary mathematics. *American Educator*. Retrieved from http://www.aft.org/pubs-reports/american_educator/fall99/amed1.pdf

Ball, D. L. (1988a). *The subject matter preparation of prospective mathematics teachers: Challenging the myths* (Research Report 88-1). East Lansing, MI: Michigan State University, National Center for Research on Teacher Education.

Ball, D. L. (1988b). *Unlearning to teach mathematics* (Issue Paper 88-1). East Lansing, MI: Michigan State University, National Center for Research on Teacher Education.

Ball, D. L., and Bass, H. (2000). Making believe: The collective construction of public mathematical knowledge in the elementary classroom. In D. C. Phillips (ed.), *Constructivism in education: Opinions and second opinions on controversial issues: Ninety-ninth yearbook of the National Society for the Study of Education* (Part 1) (pp. 193–224). Chicago: The National Society for the Study of Education.

Ball, D. L., Thames, M. H., and Phelps, G. (2008). Content knowledge for teaching: What makes it special? *Journal of Teacher Education, 59*(5), 389–407.

Ball, D. and Wilson, S. (2012). A dialogue about the professional learning continuum. Plenary presentation at the 34th Annual Meeting of the North American Chapter of the International Group for the Psychology of Mathematics Education. Kalamazoo, MI: Western Michigan University.

Baumert, J., Kunter, M., Blum, W., Brunner, M., Voss, T., Jordan, A., Klusmann, U., Krauss, S., Neubrand, M., and Tsai, Y. (2010). Teachers' mathematical knowledge, cognitive activation in the classroom, and student progress. *American Educational Research Journal, 47*(1), 133–180.

Bay-Williams, J. M., and Meyer, M. R. (2005). Why not just tell students how to solve the problem? *Mathematics Teaching in the Middle School, 10*, 340–341.

Beaugris, L. M. (2013). Mathematical observations: The genesis of mathematical discovery in the classroom. *For the Learning of Mathematics, 33*(1), 21–26.

Boaler, J., and Humphreys, C. (2005). *Connecting mathematical ideas: Middle school video cases to support teaching and learning*. Portsmouth, NH: Heinemann.

Davis, B., and Simmt, E. (2006). Mathematics-for-teaching: An ongoing investigation of the mathematics that teachers (need to) know. *Educational Studies in Mathematics, 61*, 293–319.

Doerr, H. and Lesh, R. (2003). A modeling perspective on teacher development. In R. Lesh and H. Doerr (eds.), *Beyond constructivism: Models and modeling perspectives on mathematics problem solving, learning, and teaching* (pp. 125–139). Mahwah, NJ: Lawrence Erlbaum Associates.

Eisenhart, M., Borko, H., Underhill, R., Brown, C., Jones, D., and Agard, P. (1993). Conceptual knowledge falls through the cracks: Complexities of learning to teach mathematics for understanding. *Journal for Research in Mathematics Education, 24*(1), 8–40.

English, L. D., Fox, J. L., and Watters, J. J. (2005). Problem posing and solving with mathematical modeling. *Teaching Children Mathematics, 12*, 156–163.

Gutierrez, A., Jaime, A., and Fortuny, J. (1991). An alternative paradigm to evaluate the acquisition of the Van Hiele Levels. *Journal for Research in Mathematics Education, 22*(3), 237–251.

Handal, B. (2003). Teachers' mathematics beliefs: A review. *The Mathematics Educator, 13*(2), 47–57.

Harel, G., and Lesh, R. (2003). Local conceptual development of proof schemes in a cooperative learning setting. In R. Lesh and H. Doerr (eds.), *Beyond constructivism: Models and modeling perspectives on mathematics problem solving, learning, and teaching* (pp. 359–382). Mahwah, NJ: Lawrence Erlbaum Associates.

Hart, L., Swars, S., Oesterle, S., and Kajander, A. (2012). Developing elementary teachers' mathematical knowledge for teaching: Building on what we know. In L. R. Van Zoest, J. J. Lo, and J. L. Kratky (eds.), *Proceedings of the 34th Annual Meeting of the North American Chapter of the International Group for the Psychology of Mathematics Education* (pp. 1214–1222). Kalamazoo, MI: Western Michigan University.

Hiebert, J., and Grouws, D. A. (2007). The effects of classroom mathematics teaching on students' learning. In F. Lester, Jr. (ed.), *Second handbook of research on mathematics teaching and learning: A project of the National Council of Teachers of Mathematics* (pp. 371–404). Charlotte, NC: Information Age Publishing.

Hill, H., and Ball, D. (2004). Learning mathematics for teaching: Results from California's mathematics professional development institutes. *Journal for Research in Mathematics Education, 35*(5), 330–351.

Hill, H., Sleep, L., Lewis, J., and Ball, D. (2007). Assessing teachers' mathematical knowledge: What knowledge matters and what evidence counts. In F. Lester, Jr. (ed.), *Second handbook of research on mathematics teaching and learning: A project of the National Council of Teachers of Mathematics* (pp. 111–156). Charlotte, NC: Information Age Publishing.

Holm, J., and Kajander, A. (2011). "I finally get it": Developing mathematical understanding during teacher education. *International Journal of Mathematical Education in Science and Technology, 43*(5), 563–574.

Holm, J., and Kajander, A. (2012). Interconnections of knowledge and beliefs in teaching mathematics. *Canadian Journal of Science, Mathematics, and Technology Education, 12*(1), 7–21.

Kahan, J. A., Cooper, D. A., and Bethea, K. A. (2003). The role of mathematics teachers' content knowledge in their teaching: A framework for research applied to a study of student teachers. *Journal of Mathematics Teacher Education, 6*(3), 223–252.

Kajander, A. (2010a). Elementary mathematics teacher preparation in an era of reform: The assessment and development of mathematics for teaching. *Canadian Journal of Education, 33*(1), 228–255.

Kajander, A. (2010b). Teachers constructing concepts of mathematics for teaching and learning: "It's like the roots beneath the surface, not a bigger garden." *Canadian Journal of Science, Mathematics and Technology Education, 10*(2), 87–102.

Kajander, A., Fredrickson, E., Casasola, M. and Boland, T. (2013). "Does anyone have another way?": Patterning, algebra, and inquiry in the elementary classroom. *OAME/AOEM Gazette, 51*(3), 28–34.

Kajander, A., and Holm, J. (2013). Preservice teachers' mathematical understanding: Searching for differences based on school curriculum background. *Fields Mathematics Education Journal, 1*(1), 3–20.

Kajander, A., and Jarvis, D. (2009). Mathematics for elementary teaching: Report of the working group of elementary mathematics for teaching. *Canadian Mathematics Education Forum*. Vancouver: Simon Fraser University. Retrieved from http://math.ca/Events/CMEF2009/reports/wg2-report.pdf

Kajander, A., Mason, R., Taylor, P., Doolittle, E., Boland, T., Jarvis, D., and Maciejewski, W. (2010). Nourishing the roots or staking the plants: A conversation about multiple visions of teachers' understandings of mathematics. *For the Learning of Mathematics, 30*(3), 50–56.

Kamii, C. (2004). *Young children continue to reinvent arithmetic, 2nd grade: Implications of Piaget's theory.* New York: Teachers College Press.

Lesh, R., and Doerr, H. (2003). *Beyond constructivism: Models and modeling perspectives on mathematics problem solving, learning, and teaching.* Mahwah, NJ: Lawrence Erlbaum Associates.

Ma, L. (1999). *Knowing and teaching elementary mathematics.* Mahwah, NJ: Lawrence Erlbaum Associates.

McNeal, B., and Simon, M. A. (2000). Mathematics culture clash: Negotiating new classroom norms with prospective teachers. *Journal of Mathematical Behavior, 18*(4), 475–509.

Moreira, P. C., and David, M. M. (2008). Academic mathematics and mathematical knowledge needed in school teaching practice: Some conflicting elements. *Journal of Mathematics Teacher Education, 11*(1), 23–40.

National Council of Teachers of Mathematics (NCTM). (1989). *Curriculum and evaluation standards for school mathematics.* Reston, VA: The National Council of Teachers of Mathematics, Inc.

National Council of Teachers of Mathematics (NCTM). (2000). *Principles and standards for school mathematics.* Reston, VA: The National Council of Teachers of Mathematics, Inc.

National Council on Teacher Quality (NCTQ). (2008). *No common denominator: The preparation of elementary teachers in mathematics by America's education schools.* Retrieved from http://www.nctq.org

National Mathematics Advisory Panel (NMAP). (2008). *Foundations for success: The final report of the National Mathematics Advisory Panel.* Retrieved from http://www2.ed.gov/about/bdscomm/list/mathpanel/report/final-report.pdf

Palincsar, A. S. (1998). Social constructivist perspectives on teaching and learning. *Annual Review of Psychology, 49*, 345–375.

Reid, D. A. (2011). Understanding proof and transforming teaching. Plenary presentation at the Annual Meeting of the North American Chapter of the International Group for the Psychology of Mathematics Education. Reno, NV: University of Nevada.

Silverman, J., and Thompson, P. W. (2008). Toward a framework for the development of mathematical knowledge for teaching. *Journal of Mathematics Teacher Education, 11*, 499–511.

Skemp, R. (1986). *The psychology of learning mathematics*. New York: Penguin.

Sowder, J. (2007). The mathematical education and development of teachers. In F. Lester, Jr. (ed.), *Second Handbook of Research on Mathematics Teaching and Learning* (pp. 157–224). Charlotte, NC: Information Age Publishing.

Stein, M., Remillard, J., and Smith, M. (2007). How curriculum influences student learning. In F. Lester, Jr. (ed.), *Second handbook of research on mathematics teaching and learning: A project of the National Council of Teachers of Mathematics* (pp. 319–369). Charlotte, NC: Information Age Publishing.

Stipek, D., Givvin, K., Salmon, J., and MacGyvers, V. (2001). Teachers' beliefs and practices related to mathematics instruction. *Teaching and Teacher Education, 17*(2), 213–226.

Stylianides, A. J., and Ball, D. L. (2008). Understanding and describing mathematical knowledge for teaching: Knowledge about proof for engaging students in the activity of proving. *Journal of Mathematics Teacher Education, 11*(4), 307–332.

Thompson, A. (1992). Teachers' beliefs and conceptions: A synthesis of the research. In D. A. Grouws (ed.), *Handbook of Research on Mathematics Teaching and Learning* (pp. 189–236). Hillsdale, NJ: Lawrence Erlbaum Associates.

Weiss, M., and Moore-Russo, D. (2012). Thinking like a mathematician. *Mathematics Teacher, 106*(4), 269–273.

Weller, K., Arnon, I., and Dubinsky, E. (2009). Preservice teachers' understanding of the relationship between a fraction or integer and its decimal expansion. *Canadian Journal of Science, Mathematics and Technology Education, 9*(1), 5–28.

Zack, V., and Reid, D. A. (2003). Good-enough understanding: Theorising about the learning of complex ideas (part 1). *For the Learning of Mathematics, 23*(3), 43–50.

Zack, V., and Reid, D. A. (2004). Good-enough understanding: Theorising about the learning of complex ideas (part 2). *For the Learning of Mathematics, 24*(1), 25–28.

Glossary

addition: Addition is finding the total, or sum, by combining two or more numbers.

associative property: The associative property is the idea that we can group multiplication and division operations and perform them in pairs in an order other than left to right.

circumference: Circumference is the term for the perimeter of a circle.

closed shape: A closed shape has edges that meet or a boundary with no gaps, such as a circle.

common denominator: Fractions that have the same denominator are said to have common denominators.

common factor: A common factor is a factor of two or more numbers. The *greatest* common factor is the largest of these factors.

commutative: A property of some operations that makes it possible to perform the operation in any order and get the same answer. For example, the commutative property of multiplication means that 6×4 and 4×6 result in the same product. Similarly, the commutative property of addition means that $6 + 4$ results in the same sum as $4 + 6$.

composite number: A composite number is divisible by numbers other than 1 and itself.

concave: A three-dimensional concave object has faces that meet with interior angles of less than 180 degrees; for example, a soccer ball is concave, while a three-dimensional star is not.

conjecture: A conjecture is an idea or theory proposed as a mathematical truth.

constant: A constant is a name given to an unchanging amount; for example, the number 5 is a constant.

continuous data: Continuous data involves numbers with no spaces or gaps between them. When working with continuous data, it is always possible to find another value between two numbers.

core: The core of a repeating pattern is the part or attribute that is repeating.

degree: A degree is a unit for measuring the angle of rotation. A complete rotation, such as a clock hand rotating around the face of the clock exactly once, is measured as 360 degrees. Half a rotation is thus 180 degrees, and so on.

denominator: The denominator is the bottom number in the symbolic notation of a fraction. It tells us how many parts there are in one whole.

dependent variable: A dependent variable depends on another variable to take its value. For example, in the relationship $y = 2 + x$, the value of y depends on the value used for x.

diameter: The diameter of a circle is any straight line that passes through the centre of the circle, with end points on the circumference. It is two radii that meet in a straight line.

discrete data: Discrete data is numeric data that cannot be subdivided beyond a certain point. There are gaps between the numbers.

dividend: A number or quantity into which another number or quantity (the divisor) is divided.

division: Division is a mathematical operation whereby the quotient, or result, of dividing one number or quantity by another number or quantity is determined.

divisor: A number or quantity to be divided into another number or quantity (the dividend).

domain: The domain is the set or list of allowable values that can be used as unknowns or variables in a given expression. For example, the domain might be restricted to whole numbers (and may not include fractions or decimals) for a given context.

edge: Edges are straight lines defining the outside of a closed two-dimensional shape. They are sometimes informally referred to as sides. An edge of a three-dimensional object is a line at which two faces meet.

equation: An equation is a statement of equality between two expressions. It contains an equal sign.

equivalent: Equivalent means having the same value. For example, $\frac{1}{2}$ and $\frac{2}{4}$ are equivalent fractions; 2×3 and 6×1 are equivalent expressions.

equivalent shapes: Shapes that are equivalent are the same in every way except location.

face: A face of a three-dimensional object is a flat surface on its outer surface. A cube, for example, has six faces.

factor: A factor of a number is a number that divides into that number with no remainder. Factors may be restricted to whole numbers.

first difference: The first difference is the value added (or subtracted) each time according to the recursive pattern rule. It is the difference between output values, when the input increases by 1.

function rule: The function rule in a relationship tells how to find the output for a particular input value or term number.

imaginary number: An imaginary number is a number that has a square root that is less than zero. For example, $\sqrt{-25}$ is an imaginary number and its square is 5i.

improper fraction: An improper fraction is a fraction in the form $\frac{a}{b}$ that has a value that is greater than one whole; for example, $\frac{6}{4}$ and $\frac{3}{2}$.

integers: The integers are the set of numbers that include the whole numbers together with numbers that move in the opposite direction from zero on the number line, such as $-1, -2$, and so on.

inverse operations: Inverse operations are operations that undo or reverse each other, such as addition and subtraction.

irrational number: An irrational number is a real number that cannot be written as a simple fraction (i.e., in the form $\frac{a}{b}$, where a and b are integers and b is not zero). See *rational number*.

line: A line is a one-dimensional object that extends infinitely in both directions.

line segment: A line segment has a measurable length; it is like a line, but with a definite beginning and end point.

lowest common denominator: Fractions that have the same denominator are said to have common denominators. The smallest possible such denominator for two or more fractions is called the lowest common denominator.

lowest common multiple: The lowest common multiple of two or more values is the first or smallest value that is a multiple of each; for example, the lowest common multiple of 4 and 6 is 12.

lowest terms: Fractions in lowest terms are fractions that cannot be represented in larger pieces; for example, $\frac{3}{2}$ or $\frac{1}{5}$. In this form of fraction, the numerator and denominator have no factor in common except 1.

mean: The mean is a measure of central tendency calculated by adding all the values or observations and dividing by the number of values or observations. The mean is often referred to as the *average*.

measurement model of division: The measurement model of division is enacted by counting or measuring out portions through repeated subtraction.

median: The median is a measure of central tendency that represents the middle value or observation (or midpoint between two values) in a list of values or observations ordered by size.

mixed number: A mixed number contains a whole number and a fraction; for example, $1\frac{2}{4}$ and $1\frac{1}{2}$.

mode: The mode is a measure of central tendency that gives the most common or frequent value or observation in a list of data.

multiple: A multiple of a number is that number multiplied by any other whole number.

multiplication: Multiplication is a mathematical operation whereby the product or result of combining a number or quantity with itself a given number of times is determined.

net: The net of a three-dimensional shape is a two-dimensional shape that can be folded along its edges to create the three-dimensional object.

non-linear: A non-linear relationship is represented by a graph that is not a straight line.

numerator: The numerator is the top number in the symbolic notation of a fraction. It tells us how many pieces we are talking about. The size of the pieces is determined by the denominator. See *denominator*.

parallel: Parallel refers to surfaces (or lines) that do not ever meet (at least in Euclidean geometry).

parallelogram: A parallelogram is a four-sided shape with both sets of opposite sides equal and parallel. A parallelogram may or may not be rectangular (having 90-degree or right-angled corners).

partitive model of division: The partitive model of division is enacted by splitting a quantity into equal groups.

perimeter: The perimeter of a shape is its boundary or the measure of its boundary.

perpendicular line: A perpendicular line or line segment is one that meets another line or segment at a 90-degree angle, sometimes called a *right angle*.

place value: Place value refers to the value of an individual digit as determined by its position in a number. For example, in the number 497, the value of the 4 is actually 4 × 100 (or 400), the value of the 9 is actually 9 × 10 (or 90), and the value of the 7 is actually 7 × 1 (or 7).

plane: A plane is a flat or two-dimensional surface.

polygon: A polygon is a two-dimensional shape with straight lines as sides or edges. A regular polygon is a polygon with edges that are all equal in length. A square is a regular polygon, and a (non-square) rectangle is a polygon.

prime factor: A factor that has no factors other than 1 and itself.

prime number: A prime number is a whole number that is divisible by only the whole numbers 1 and itself, for example, 7.

prism: A prism is a three-dimensional shape that has at least two parallel faces that are the same.

product: A product is the answer to a multiplication problem.

proper fraction: A proper fraction is a fraction in the form $\frac{a}{b}$ that has a value that is less than one whole; for example, $\frac{3}{4}$.

proportional shapes: Similar objects are different sizes, but all angle measures are preserved. Mathematically, we say the shapes are proportional. If two shapes are proportional, then respective sides are in the same ratio. See *similar objects*.

quotient: A quotient is the answer to a division question. It is sometimes used to refer to the whole number part of the answer.

radius: The radius is the distance from the centre of a circle to any point on its circumference.

range: The range of a calculation or expression is the possible or allowable set of resultant values. For example, if a problem was about someone's age, then the range would include only positive values, although fractions could be allowed.

ratio: A ratio is a comparison of two quantities of the same unit. It can be expressed as a comparison (3:4) or a fraction ($\frac{3}{4}$).

rational number: A rational number can be written as a simple fraction, in the form $\frac{a}{b}$, where a and b are integers and b is not zero. See *irrational number*.

reciprocals: Reciprocals are rational numbers with the relationship $\frac{a}{b}$ and $\frac{b}{a}$, where a and b are non-zero.

recursive rule/recursive solution: The recursive rule or solution for a relationship tells you how much the output changes when the input increases by 1, from a given starting point; for example, *start with 2 and add 3 each time*.

reflective symmetry: Reflective symmetry means that a line can be drawn through a shape that acts like a mirror, and the shape will be identical on either side of the line.

remainder: The remainder is what is left when we divide to the nearest whole number quotient.

right angle: A right angle is a 90-degree angle, or one-quarter of a full rotation. Perpendicular lines meet at a 90-degree angle.

similar objects: Similar objects are different sizes, but all angle measures are preserved. Mathematically, we say the shapes are proportional. See *proportional shapes*.

square root: The square root of a number is the number that is multiplied by itself to result in the original value. For example, 20 is the square root of 400 because 20 × 20 is 400.

subtraction: Subtraction is the mathematical operation whereby the difference between two numbers or quantities is determined. It can also be used to measure how far apart two numbers are.

sum: A sum is the result of combining or adding quantities.

supplementary angles: Two angles that add up to 180 degrees are called supplementary.

surface area: The surface area of a three-dimensional object refers to the total (two- dimensional) area of its outside surfaces.

symmetry: Symmetry is a geometric property of being balanced about a line or a point (or a plane in three dimensions).

tessellation: A tessellation is a shape that can completely cover a flat surface with no gaps or overlap.

transversal: A transversal is a straight line or segment that crosses or intersects two or more parallel lines.

unit fraction: A unit fraction is a fraction with 1 as the numerator. It represents one piece of the size of the fraction unit specified by the denominator. For example, $\frac{1}{4}$ is one piece of a whole that is made up of 4 one-fourths.

variable: A variable is a symbol that represents a value that can vary.

vertex: A vertex is the point at which two or more edges meet; we might informally call it a *corner*. The plural of vertex is vertices; a hexagon has 6 vertices.

zero pair: One positive and one negative number of the same value is a zero pair; for example, 1 + (–1), which is another way to represent zero. Zero pairs are necessary for some models of integer operations.